GRADE 9

STP MATHEMATICS
for Jamaica

Workbook

S Chandler • E Smith

Nelson Thornes

Published in 2012 by:
Nelson Thornes Ltd
Delta Place
27 Bath Road
CHELTENHAM
GL53 7TH
United Kingdom

12 13 14 15 16 / 10 9 8 7 6 5 4 3 2 1

A catalogue record for this book is available from the British Library

ISBN 978 1 4085 1804 5

Page make-up and illustrations by Pantek Media, Maidstone, Kent

Printed in China

Contents

1 Working with Integers

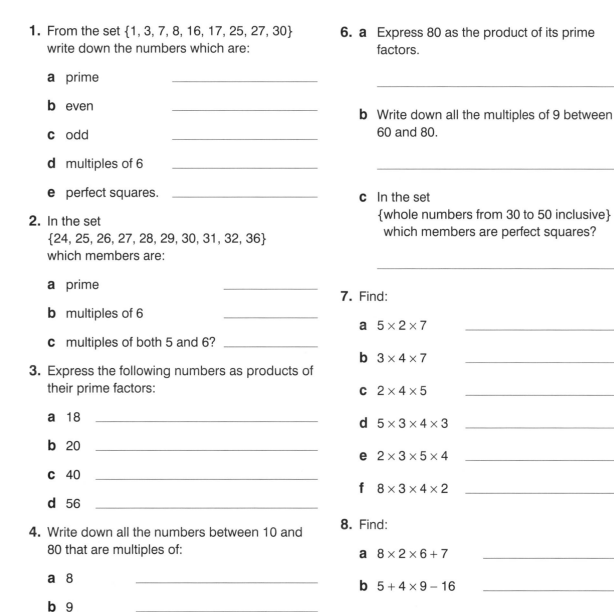

1. From the set {1, 3, 7, 8, 16, 17, 25, 27, 30} write down the numbers which are:

   a prime _____

   b even _____

   c odd _____

   d multiples of 6 _____

   e perfect squares. _____

2. In the set {24, 25, 26, 27, 28, 29, 30, 31, 32, 36} which members are:

   a prime _____

   b multiples of 6 _____

   c multiples of both 5 and 6? _____

3. Express the following numbers as products of their prime factors:

   a 18 _____

   b 20 _____

   c 40 _____

   d 56 _____

4. Write down all the numbers between 10 and 80 that are multiples of:

   a 8 _____

   b 9 _____

   c 7 and 11. _____

5. Find the lowest number that is a multiple of:

   a both 3 and 5 _____

   b both 3 and 6 _____

   c both 5 and 6. _____

6. **a** Express 80 as the product of its prime factors.

   _____

   b Write down all the multiples of 9 between 60 and 80.

   _____

   c In the set {whole numbers from 30 to 50 inclusive} which members are perfect squares?

   _____

7. Find:

   a $5 \times 2 \times 7$ _____

   b $3 \times 4 \times 7$ _____

   c $2 \times 4 \times 5$ _____

   d $5 \times 3 \times 4 \times 3$ _____

   e $2 \times 3 \times 5 \times 4$ _____

   f $8 \times 3 \times 4 \times 2$ _____

8. Find:

   a $8 \times 2 \times 6 + 7$ _____

   b $5 + 4 \times 9 - 16$ _____

   c $350 \div 7 + 4 \times 4$ _____

   d $10 \div 2 + 6 \times 3$ _____

   e $16 - 45 \div 3 + 12$ _____

   f $9 \times 6 - 6 \times 5$ _____

   g $63 \div 9 + 35 \div 5$ _____

9. Insert > or < between each pair of numbers:

a 5 _____ –4

b –7 _____ –6

c 0 _____ –2

d 8 _____ –7

e –1 _____ –6

f 3 _____ –2

10. Find, using a number line if it helps:

a –4 + (–2) _____

b –7 – (–4) _____

c –5 – (+5) _____

d +8 + (–4) _____

e +9 – (–3) _____

f +10 – (+4) _____

11. Calculate

a 5 – 2 + 6 _____

b 12 – (–9) – 2 _____

c 3 – 5 – (–4) _____

d –6 + 3 – 5 _____

e 7 – 3 – 2 _____

f 8 + (–3) – (–5) _____

12. Find

a $(4) \times (-3)$ _____

b $(-7) \times (-4)$ _____

c $(+12) \div (-3)$ _____

d $(-6) \div (-2)$ _____

e $(-5)^2$ _____

f $(-8) \times 2$ _____

1. 3 and 4 are members of $\mathbb{N}$. Which of the following results are members of $\mathbb{N}$?

Explain your answers.

a $3 + 4$ **b** 3×4

c $3 \div 4$ **d** $4 - 3$

e $3 - 4$

_____

_____

2. $3, -4, \dfrac{1}{2}, \dfrac{\pi}{2}$

Choose one number from this list to complete the following statements. Use each number once only.

a _____ is included in $\mathbb{N}$

b _____ is included in $\mathbb{R}$

c _____ is included in $\mathbb{Z}$

d _____ is included in $\mathbb{Q}$

3. Find the LCM of the following sets of numbers:

a 10, 20 _____

b 4, 5, 6 _____

c 6, 8, 12 _____

d 3, 5, 9 _____

4. Find:

a $\dfrac{1}{3} + \dfrac{1}{4}$ _____

b $\dfrac{1}{2} + \dfrac{3}{4}$ _____

c $\dfrac{7}{12} + \dfrac{5}{8}$ _____

d $\dfrac{2}{3} + \dfrac{8}{21} + \dfrac{2}{7}$ _____

5. Find:

a $\dfrac{3}{4} - \dfrac{1}{2}$ _____

b $\dfrac{5}{6} - \dfrac{2}{3}$ _____

c $\dfrac{5}{8} - \dfrac{3}{5}$ _____

d $\dfrac{11}{15} - \dfrac{2}{3}$ _____

6. Find:

a $2\dfrac{1}{5} + 1\dfrac{5}{8}$ _____

b $5\dfrac{3}{7} - 4\dfrac{1}{5}$ _____

c $3\dfrac{3}{4} - 1\dfrac{5}{6} + 2\dfrac{1}{3}$ _____

d $1\dfrac{7}{10} - \dfrac{8}{9} + 3\dfrac{1}{5}$ _____

e $5\dfrac{1}{4} + 2\dfrac{2}{3} - 7\dfrac{1}{12}$ _____

7. Find:

a $\dfrac{3}{4} \times \dfrac{7}{15}$ _____

b $5 \times \dfrac{3}{7}$ _____

c $\dfrac{3}{4} \times 1\dfrac{1}{6}$ _____

d $\dfrac{5}{6} \times \dfrac{3}{10} \times \dfrac{7}{12}$ _____

8. Find the missing numbers:

a $\dfrac{2}{3} \times \dfrac{\square}{4} = \dfrac{1}{2}$ _____

b $\dfrac{\square}{\square} \times \dfrac{5}{12} = 1$ _____

c $\dfrac{7}{15} \times \dfrac{\square}{21} = \dfrac{1}{9}$ _____

9. Write down the reciprocal of:

a 6 _____

b $\dfrac{2}{3}$ _____

c $\dfrac{5}{9}$ _____

d 50 _____

10. Find:

a $\frac{1}{2} \div \frac{1}{3}$ _____

b $3\frac{1}{3} \div 4$ _____

c $\frac{5}{7} \div 10$ _____

d $5 \div \frac{3}{5}$ _____

11. Find:

a $4\frac{7}{8} \div 1\frac{1}{4}$ _____

b $1\frac{6}{7} \div 6\frac{1}{2}$ _____

c $5\frac{1}{3} \times 2\frac{5}{8} \div 4\frac{2}{3}$ _____

d $1\frac{4}{5} \times 2\frac{1}{7} \div \frac{3}{7}$ _____

12. Find:

a $\left(\frac{4}{7} + \frac{3}{5}\right) \div 8\frac{1}{5}$ _____

b $\dfrac{2\frac{1}{2}}{1\frac{1}{3} - \frac{1}{3}}$ _____

c $4\frac{3}{5} - \left(1\frac{7}{10} \times 2\frac{1}{2}\right)$ _____

d $\left(1\frac{1}{2} - \frac{2}{5}\right) \times \left(\frac{2}{3} + \frac{1}{6}\right)$ _____

13. Find:

a $\frac{2}{7} + \frac{1}{2} \div 1\frac{1}{3}$ _____

b $\left(\frac{2}{5} + \frac{3}{7}\right) \div 8\frac{6}{7}$ _____

c $\left(2\frac{3}{10} \times 2\frac{1}{2}\right) - 5\frac{2}{5}$ _____

d $\left(2\frac{1}{4} - \frac{3}{5}\right) \times \left(\frac{1}{3} + \frac{5}{6}\right)$ _____

14. Express the following decimals as fractions in their lowest terms:

a 0.16 _____

b 3.15 _____

c 0.44 _____

d 0.025 _____

15. Express these fractions as decimals:

a $\frac{7}{8}$ _____

b $\frac{3}{16}$ _____

c $\frac{9}{32}$ _____

d $\frac{7}{40}$ _____

e $2\frac{5}{8}$ _____

f $3\frac{15}{32}$ _____

16. Use the dot notation to write these fractions as decimals:

a $\frac{3}{7}$ _____

b $\frac{9}{11}$ _____

c $\frac{2}{13}$ _____

17. Find, without using a calculator:

a $1.46 + 5.72$

_____

b $0.0279 - 0.0083$

_____

c 37×0.07

_____

d $5.04 \div 0.9$

_____

e $300 \times 0.2 - 10.4$

_____

f $(2.9 - 0.5) \div 0.8$

_____

g $(1.5 - 0.3) \div 0.6$

_____

18. Find, without using a calculator:

a $5.92 - 3.66$

_____

b $0.0876 + 1.042$

_____

c 73×0.04

_____

d $0.112 \div 0.07$

_____

e $(4.7 - 1.1) \div 0.9$

_____

f $3.6 + 1.4 \times 0.2$

_____

19. Evaluate:

a $2.4(0.7 - 0.08) + 8.75$

_____

b $(40 \times 0.03) \div (4.3 - 3.1)$

_____

c $\dfrac{0.02 \times 2.04}{1.7 \times 0.12}$

_____

d $\dfrac{3.96 \times 1.2}{0.6 \times 3}$

_____

20. Place > or < between the following pairs of numbers:

a $\dfrac{5}{11}$ _____ 0.5

b 0.23 _____ $\dfrac{2}{9}$

c $\dfrac{3}{13}$ _____ 0.22

21. Arrange the following numbers in ascending order of size:

a $\dfrac{4}{7}, 0.57, \dfrac{5}{9}$ _____

b $\dfrac{7}{20}, 0.36, \dfrac{8}{25}, 0.33$ _____

c $\dfrac{4}{9}, 0.51, \dfrac{5}{12}, 0.42$ _____

22. Arrange the following numbers in descending order of size:

a $0.58, \dfrac{4}{7}, \dfrac{5}{9}$ _____

b $\dfrac{1}{6}, \dfrac{1}{7}, 0.14$ _____

c $\dfrac{12}{13}, 0.08, \dfrac{11}{15}, \dfrac{13}{14}$ _____

23. Find the value of:

a $4^2 \times 5^2$ _____

b 2×3^2 _____

c 5.92×10^3 _____

d 2.1×10^2 _____

24. Find the value of:

a 3^4 _____

b $6^2 \times 2^3$ _____

c $(2^2)^3$ _____

d 2^5 _____

25. Write as a single number in index form:

a $2^2 \times 2^5$ _____

b $a^3 \times a^4$ _____

c $3^5 \div 3^2$ _____

d $a^8 \div a^2$ _____

26. Write as a single expression in index form

a $5^8 \div 5^3$ _____

b $2^5 \div 2^4$ _____

c $a^3 \div a^5$ _____

d $b^7 \times b^4$ _____

27. Simplify:

  a $3(a^3)^2$ _____

  b $(3a^3)^2$ _____

  c $(2a^3)^3$ _____

  d $(2a^5)^3$ _____

28. Write these numbers as ordinary numbers:

  a 1.6×10^3 _____

  b 7.02×10^2 _____

  c 4.27×10^4 _____

  d 5.08×10^6 _____

29. Write the following numbers in standard form:

  a 425 _____

  b 5200 _____

  c 75 640 _____

  d 81 000 000 _____

  e 36 000 _____

  f 621 000 _____

30. Give each number correct to

    i 3 decimal places

    ii 3 s.f.

  a 0.049896

    i _____   **ii** _____

  b 3.0747

    i _____   **ii** _____

  c 0.008734

    i _____   **ii** _____

  d 67.09046

    i _____   **ii** _____

  e 469.6037

    i _____   **ii** _____

31. **i** Give a rough value for each calculation.

    ii Now use a calculator to give the answer correct to 3 s.f.

  a 0.747×1.0043

    i _____   **ii** _____

  b $(0.47)^3$

    i _____   **ii** _____

  c 530×97.4

    i _____   **ii** _____

  d $\dfrac{0.028 \times 4.14}{3.74}$

    i _____   **ii** _____

32. Illustrate each range on a number line:

  a $3 \leqslant x \leqslant 7$

  _____

  b $-4 \leqslant x < 9$

  _____

  c $-1 < x \leqslant 10$

  _____

  d $0 < x < 4.7$

  _____

  e $0 \leqslant x < 0.7$

  _____

  f $0.25 \leqslant x \leqslant 0.5$

  _____

  g $1.5 \leqslant x \leqslant 2.7$

  _____

33. Illustrate on a number line the range of possible values for each of the following corrected numbers. They are given correct to 1 d.p.

a 2.5

_____

b 0.7

_____

c 5.9

_____

d 0.2

_____

e 15.4

_____

34. Illustrate on a number line the range of possible values for each of the following corrected numbers. They are given correct to 2 d.p.

a 0.46

_____

b 1.29

_____

c 5.37

_____

d 18.52

_____

e 82.65

_____

35. Usain's income for the week, correct to the nearest $1000, was $35 000. Find the range in which his income lies.

_____

36. The attendance at a test match was 27 000, correct to the nearest 1000. What is the difference between the largest number that could have been present and the smallest number?

_____

37. Correct to the nearest square metre, the area of a room is 85 m². Find the range in which the area of this room lies.

_____

38. Daisy asks a friend to make her a drawing board. The measurements she gives are 60 cm by 40 cm, each to the nearest centimetre. Find the range of values in which each actual measurement lies.

_____

39. Milly collects stamps. She describes the number of stamps she has as 120, correct to the nearest 5. What is

a the maximum number of stamps Milly has?

_____

b the minimum number of stamps she has?

_____

40. A sheet of metal must be 1.50 mm thick, correct to the nearest $\frac{1}{100}$ th of a millimetre. Find the range of values in which this thickness lies.

_____

3 Equations and Formulae

1. Simplify:

   a $2x + y - 3x$ _____

   b $5p + 2p$ _____

   c $3a - (-4a)$ _____

   d $4p + 2q - (-3p)$ _____

   e $p - (-3q)$ _____

   f $x + 3y - (-2x)$ _____

2. Simplify:

   a $2p \times 3p$ _____

   b $4a \times 5b$ _____

   c $12p \div 3p$ _____

   d $a \times (-b)$ _____

   e $12b \div (-3b)$ _____

   f $(-8x) \div (-2x)$ _____

3. Simplify:

   a $x - 4(x - y)$ _____

   b $3(a + b) + 2(a - b)$ _____

   c $5(p + q) - 3(p - q)$ _____

   d $4x - (x - y)$ _____

   e $5y - 3(y + z)$ _____

4. Solve the following equations:

   a $2x + 7 = 3x + 1$ _____

   b $11 - 2a = 5 + a$ _____

   c $7 + x = 13 - 5x$ _____

   d $4z + 9 = 7z + 18$ _____

   e $3 + 2a = 5a - 6$ _____

   f $8p + 3 = 5p - 4$ _____

5. Solve the following equations:

   a $3x - 1 = 3 - 3(x + 2)$ _____

   b $3y = 8 - 2(y - 1)$ _____

   c $5 - 2(3 - 2a) = 6$ _____

   d $6x = 3x - (x - 8)$ _____

   e $4(3p - 2) - 5(3p - 1) = 0$ _____

   f $7x = x - (x - 16)$ _____

6. Write down a formula connecting the given letters:

   a A number N is equal to twice the sum of two numbers a and b.

   _____

   b A number P is equal to twice the product of two numbers q and r.

   _____

   c A number X is equal to 3 times the square of a number y.

   _____

   d A number Y is equal to a number x plus its square.

   _____

   e Screws are sold at $\$x$ per 100 g. What would be the cost, $\$C$, of 1 kg of the same screws?

   _____

7. a Given that $t = \dfrac{v - u}{f}$ find t when $v = 10$, $u = 3$ and $f = 14$.

   _____

   b If $p = qt - 2r$, find p when $q = 4$, $t = -3$ and $r = 2$.

   _____

c If $s = \frac{1}{2}(a + b + c)$ find s when $a = 1.4$, $b = 2.5$ and $c = 3.3$.

_____

8. a $x = (y - 2z)^2$
Find x when $y = 5$ and $z = -2$.

_____

b Given that $A = \frac{1}{2}(ab + bc + ca)$ find A when $a = 3$, $b = \frac{1}{2}$ and $c = 8$.

_____

c If $p = \frac{q - 2r}{q - r}$ find p when $q = 5$ and $r = -\frac{1}{2}$

_____

9. Make the letter in brackets the subject of the formula.

a $a = b + c$ (c) _____

b $P = 3q + t$ (q) _____

c $a = b + 2c$ (b) _____

d $4x = y - z$ (z) _____

e $R = P - Q$ (P) _____

10. Make the letter in brackets the subject of the formula.

a $X = Y - Z$ (Y) _____

b $P = q + t$ (t) _____

c $a = \frac{b}{2c}$ (b) _____

d $3x = yz$ (z) _____

e $R = P + Q$ (P) _____

11. Make the letter in brackets the subject of the formula.

a $a = 2b - c$ (b) _____

b $z = x + 3y$ (y) _____

c $r = 4p - q$ (p) _____

d $3d - e = \frac{c}{4}$ (d) _____

e $p - q = \frac{r}{3}$ (r) _____

f $\frac{a}{2} - 4 = 3b - c$ (b) _____

12. a Make b the subject of the formula $a = \frac{b}{c}$

_____

b Find a when $b = 10$ and $c = \frac{1}{2}$. _____

c Find c when $a = -\frac{1}{2}$ and $b = 8$. _____

13. A number x is equal to the sum of twice a number y and half of a number z.

a Find a formula for x in terms of y and z.

_____

b Find x when $y = -\frac{1}{4}$ and $z = 3$. _____

c Make y the subject of this formula.

_____

14. A number a is equal to twice the square of a number b plus a number c.

a Find a formula for a in terms of b and c.

_____

b Find a when $b = -3$ and $c = 4$.

_____

c Make c the subject of this formula.

_____

15. Given that $I = \frac{PRT}{100}$

a Find P in terms of the other letters.

_____

b Find R in terms of the other letters.

_____

c Find P when $I = 100$, $R = 5$ and $T = 4$.

_____

d Find R when $P = 7000$, $I = 875$ and $T = 5$.

_____

4 Percentages

1. a Express $\frac{4}{5}$ as

    i a decimal _____

    ii a percentage. _____

  b Express 72% as

    i a decimal _____

    ii a fraction. _____

  c Express 0.65 as

    i a percentage _____

    ii a fraction. _____

2. Express the first quantity as a percentage of the second:

  a 50 cm, 4 m _____

  b 210 cm², 350 cm² _____

  c 58 cm³, 464 cm³ _____

  d 75 cm, 2 m _____

  e 4950 cℓ, 3 litres. _____

3. Find the value of:

  a 45% of 40 _____

  b $12\frac{1}{2}$% of 288 km _____

  c 78% of $95 000 _____

  d $66\frac{2}{3}$% of 5.22 cm. _____

4. Find the percentage error for each of the following values:

  a measured length 154.5 cm, actual length 150 cm.

    _____

  b measured area 561 cm², actual area 550 cm².

    _____

  c Measured mass 2.03 kg, actual mass 2 kg.

    _____

  d Estimated cost $19 240, actual cost $18 500.

    _____

5. Vilietha scored 10 marks out of 40. What percentage was this?

  _____

6. Forty-four students go on a school trip, of whom 25% have never been on a school trip before. How many students have been on a trip before?

  _____

7. a Increase:

    i 500 by 34% _____

    ii 320 by 12%. _____

  b Decrease:

    i 400 by 28% _____

    ii 250 by 12%. _____

8. Last year a factory employed 200 people. This year the workforce has been cut by 10%.

  a How many people are employed at the factory now?

    _____

  b Next year they hope to increase the present workforce by 10%. How many extra people will they take on?

    _____

9. a An article costing $2400 is sold to make a profit of $600. Find the percentage profit.

    _____

  b In a sale a chair costing the retailer $30 000 is sold at a loss of $4500. Find the percentage loss.

    _____

10. Use the given data to find the selling price:

   a cost price $5500, profit 24%

   _____

   b cost price $6400, loss $12\frac{1}{2}$%

   _____

   c cost price $14500, profit 120%

   _____

   d cost price $8500, loss 40%.

   _____

11. Find the weekly cash increase if:

   a John Baisden earning $26000 p.w. receives a rise of 10%

   _____

   b Peter Joseph earning $35000 p.w. receives a rise of 6%.

   _____

12. Assuming that the rate of general consumption tax is 15% find the purchase price of:

   a an electric kettle costing $5400 + sales tax

   _____

   b a tool set costing $12000 + sales tax.

   _____

13. Find the yearly tax due on a taxable income of $2400000 when the rate of income tax is:

   a 20% _____   **b** 25% _____

   c 30% _____   **d** 34% _____

14. a In a sale a shop offered a discount of 30% on a dress marked $14000. Find the cash price.

   _____

   b To get rid of old stock a department store offers a discount of 40% on a jacket marked $16500. Find the cash price.

   _____

15. In this question use the given data to find the cost price:

   a selling price $1280, profit 60% _____

   b selling price $1350, profit $66\frac{2}{3}$% _____

   c selling price $3024, profit 12% _____

   d selling price $16800, loss 30% _____

   e selling price $3600, loss $33\frac{1}{3}$% _____

   f selling price $2100, loss 65%. _____

16. a The purchase price of a computer game is $6720. This includes sales tax at 20%. Find the price before the sales tax was added.

   _____

   b A calculator bought for $1600 is sold at a loss of 40%. Find the selling price.

   _____

17. a Scrap metal bought for $72000 is sold at a profit of 80%. Find the selling price.

   _____

   b By selling some gold for $415800 a dealer makes a profit of 120%. What did she pay for it?

   _____

18. a Marcia's measured height is 153 cm but her actual height is 153.4 cm. Find the percentage error.

   _____

   b What is 22% of $761? _____

   c By selling a food container for $1620 a retailer makes a profit of 35%. Find the cost price.

   _____

19. Eggs are bought from a farm at $600 for a tray holding 36. They are sold at $276 per dozen. Find the percentage profit.

_____

20. Find the simple interest on:

a $30 000 invested for 2 years at 6%

_____

b $75 000 invested for 5 years at 8%

_____

c $16 680 invested for 4 years at 9%

_____

d $21 780 invested for 6 years at $7\frac{3}{4}$%.

_____

21. Find the amount if:

a $36 000 is invested for 4 years at 6% simple interest

_____

b $10 600 is invested for 5 years at 3% simple interest

_____

c $15 500 is invested for 10 years at 8% simple interest.

_____

22. a How long must $24 000 be invested at 5% simple interest to give interest of $7200?

_____

b What sum of money earns $36 480 simple interest if invested for 8 years at 8%?

_____

c Find the annual percentage rate that earns $14 760 simple interest when $82 000 is invested for $4\frac{1}{2}$ years.

_____

23. Find the compound interest on:

a $80 000 invested for 2 years at 10% p.a.

_____

b $59 000 invested for 2 years at 5% p.a.

_____

c $45 000 invested for 2 years at $3\frac{1}{2}$% p.a.

_____

d $12 000 invested for 3 years at 4% p.a.

_____

e $35 000 invested for 3 years at 7% p.a.

_____

24. a An apartment bought for $2 600 000 appreciates 5% each year. What will it be worth in 2 years' time? Give your answer correct to the nearest $1000.

_____

b A car bought for $1 800 000 depreciates by 20% each year. What will it be worth in 2 years' time?

_____

25. Novelia bought an antique silver teapot for $30 000. For the first 2 years it increased in value by 10% each year. For the next 2 years it depreciated by 10% each year. What was it worth:

a after 2 years   _____

b after 4 years?   _____

Give any answer that is not exact correct to the nearest $100.

26. Natasha decided to go on a diet. Every month she aimed to lose 2% of her weight at the beginning of that month. When she started her diet she weighed 96 kg. Assuming she manages to keep to her target, what should she weigh after 3 months?

_____

27. A flat screen is advertised at $36 000. If bought on hire purchase the terms are 30% deposit plus 18 monthly payments of $1540. Work out:

a the deposit

_____

b how much is saved by paying cash.

_____

28. A dress costs $10 800 cash, or you can pay a deposit of $2250 plus 12 monthly instalments of $806.

a How much more does the dress cost on HP compared with paying cash?

_____

b Express the additional cost as a percentage of the cost price.

_____

This table gives the equivalent of J$100 in various currencies.

Jamaican $	US $	Canadian $	Euro €	UK £	East Caribbean $
100	1.18	1.25	0.96	0.75	3.16

Use this table for questions 29 to 32.

29. Convert:

a J$40 000 into US dollars

_____

b J$80 000 into UK pounds

_____

c J$120 000 into Canadian dollars

_____

d J$5000 into euros.

_____

30. Use the table:

i to estimate

ii to calculate to the nearest whole number, the equivalent value in Jamaican dollars of

a £500

i _____   **ii** _____

b 400 US dollars

i _____   **ii** _____

c 700 Canadian dollars

i _____   **ii** _____

d 645 euros

i _____   **ii** _____

31. Use the table to convert:

a 500 euros into US dollars

_____

b £1250 into euros

_____

c US$6900 to East Caribbean dollars

_____

d 1800 euros to UK pounds.

_____

32. Liza Jardine arrived in Jamaica with US$625.

a How many Jamaican dollars did she receive when she exchanged them?

_____

b She spent J$42 500. When she left Jamaica she exchanged what remained back into US dollars. How many US$ did she receive?

_____

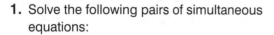

1. Solve the following pairs of simultaneous equations:

a $x + y = 7$

$3x + y = 15$ _____

b $2x + y = 7$

$5x + y = 10$ _____

c $5x + 2y = -2$

$3x + 2y = 2$ _____

d $4x + y = 10$

$7x + y = 19$ _____

2. Solve the following pairs of simultaneous equations:

a $4x + y = 16$

$3x - y = 5$ _____

b $5x + 2y = 16$

$9x - 2y = 40$ _____

c $3a + 4b = 6$

$7a - 4b = 14$ _____

d $4x + y = 11$

$5x - y = 7$ _____

3. Solve the following pairs of simultaneous equations:

a $5x + 2y = 24$

$x - 2y = 0$ _____

b $3x - 2y = 11$

$x + 2y = 1$ _____

c $4x + 3y = 14$

$2x - 3y = -11$ _____

d $a - 3b = 1$

$5a + 3b = 11$ _____

4. Solve the following pairs of simultaneous equations:

a $7x - 2y = -25$

$x - 2y = -7$ _____

b $3a - 5b = 7$

$6a + 5b = -16$ _____

c $8p - q = 10$

$4p - q = 4$ _____

d $5x - y = 14$

$2x + y = 0$ _____

5. Solve the following pairs of simultaneous equations:

a $6x + 5y = 7$

$3x + y = 2$ _____

b $7x + 3y = 13$

$x + 6y = 13$ _____

c $8a + 3b = 5$

$4a - 5b = 9$ _____

d $x + 4y = 5$

$9x - 2y = 26$ _____

6. Solve the following pairs of simultaneous equations:

a $3x + 4y = 5$

$2x + 3y = 3$ _____

b $5x - 6y = -15$

$7x + 4y = -21$ _____

c $4x - 3y = -7$

$3x + 2y = 16$ _____

d $8x - 7y = 31$

$5x + 6y = 9$ _____

7. Solve the following pairs of simultaneous equations:

a $x + y = 4$

$x + 3y = 14$ _____

b $4x + 3y = 17$

$5x - 2y = 4$ _____

c $6x + y = 30$

$3x - 4y = 15$ _____

d $x - 3y = 11$

     $3x - 2y = -2$ _____

8. Solve the following pairs of simultaneous equations:

a $y = 7 - x$

     $2x + y = 11$ _____

b $10 + x = 7y$

     $x + 2y = 8$ _____

c $x = 3 + y$

     $2x = 10 + 5y$ _____

d $2y = 2 + x$

     $y = x + 6$ _____

9. Use the substitution method to solve the following pairs of simultaneous equations:

a $3x + y = 9$

     $7x - 2y = 8$ _____

b $2x + 5y = 11$

     $6x = 5 - y$ _____

c $3a = 1 + 2b$

     $4a = 3b$ _____

10. The sum of two numbers is 26. Their difference is 12. Find the numbers.

_____

11. Three times a number added to a second number is 23. The first number added to twice the second number is 21. Find the two numbers.

_____

12. In this triangle the angle marked $x°$ is twice as large as the angle marked $y°$. Find the values of x and y.

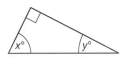

_____

13. The perimeter of this triangle is 27 cm. AC = BC and AC − AB = 3 cm. Find the values of x and y.

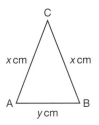

_____

14. A school photographer offers a package of 3 large photographs and 5 small photographs for $2400. As an alternative the cost of 4 large photographs and 7 small photographs is $3320. One large photograph costs $x and one small photograph costs $y. Form two equations in x and y, and solve them.

_____

Hence write down the cost of one small and one large photograph.

_____

15. A year ago a mother was 4 times as old as her son. In 15 years' time she will be twice as old as her son. Find the present age of:

a   the son _____

b   the mother. _____

16. Solve the following pairs of equations graphically. In each case, on graph paper, draw axes for x and y using values in the ranges indicated. Take 2 cm ≡ 1 unit on both axes.

a $x + y = 5$ $0 \leqslant x \leqslant 6$

     $2x - y = 1$ $-2 \leqslant y \leqslant 6$

_____

b $x + y = 4$ $-1 \leqslant x \leqslant 5$

     $y = x + 1$ $-1 \leqslant y \leqslant 6$

_____

c $x + 3y = 4$ $-2 \leqslant x \leqslant 5$

     $4x - 2y = -5$ $-1 \leqslant y \leqslant 4$

_____

6 Geometry and Construction

1. Find the size of each of the marked angles.

a

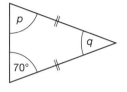

_____

b

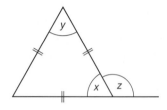

_____

c

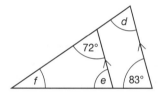

_____

_____

2.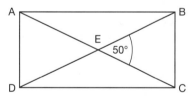

ABCD is a rectangle in which BÊC = 50°.

Find the size of EB̂C.

_____

3.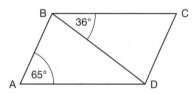

ABCD is a parallelogram.

a Find the size of BĈD _____

b Find the size of AB̂D. _____

4.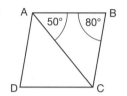

ABCD is a parallelogram in which AB̂C = 80° and BÂC = 50°. Show that ABCD is also a rhombus.

_____

_____

5.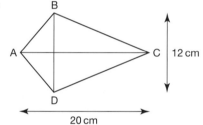

ABCD is a kite. AC = 20 cm and BD = 12 cm. Find the area of:

a the kite ABCD

_____

b triangle ABC.

_____

In questions 6 to 20 construct the given figure using only a ruler and compasses. Carry out these constructions outside your workbook.

6.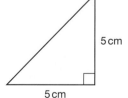

16

7.

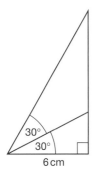

30°
30°
6 cm

8.

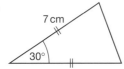

7 cm
30°

9. Construct △ABC in which AC = 10 cm, AB = 8 cm and BÂC = 60°.

10. Construct △PQR in which PR = 8.6 cm, P̂ = 45° and R̂ = 60°.

11. Construct △DEF in which DE = 5 cm, D̂ = 30° and Ê = 120°.

12. Draw a line AB 10 cm long. Construct an angle of 30° at A. Construct an angle of 60° at B. Label C, the point where the arms of A and B cross.

What should the size of angle C be?

_____

Measure Ĉ as a check.

13. Construct a square of side 7 cm. Draw the diagonals of this square. Measure and record the length of one of these diagonals.

_____

14. Construct a quadrilateral in which Â = 90°, AB = 12 cm, B̂ = 60°, AC = 5 cm and BD = 7 cm. Measure:

a the length of CD _____

b the angle D. _____

15. Construct a rhombus ABCD in which the lengths of the diagonals are 6 cm and 8 cm. Measure and record the length of one of the sides.

_____

16. Construct the isosceles triangle PQR in which PR = 7 cm and PQ = QR = 8 cm. Construct the perpendicular bisector of QR. Explain why this line is not a line of symmetry of PQR.

_____

_____

17. Construct △ABC in which AB = 7 cm, BC = 9 cm and AC = 8 cm. Construct the perpendicular bisector of AB and the perpendicular bisector of BC.

Mark E, the point where these two bisectors cross. Draw a circle with centre E and radius equal to the distance EA. This circle should pass through B and C. Measure and record its radius.

_____

18. Construct a parallelogram ABCD whose diagonals intersect at E, given that AC = 12 cm, BD = 8 cm and AÊD = 60°. Measure and write down the lengths of AB and BC.

_____

19. Construct a trapezium ABCD in which AB = 12 cm, BC = 7 cm, CD = 8 cm and AB̂C = 60°. Measure and write down the length of AD, AC and BD.

AD = _____ cm

AC = _____ cm

BD = _____ cm

20. Construct a triangle ABC in which AB = 6 cm, BC = 9 cm and AC = 10 cm. Construct the perpendicular bisector of AC and the perpendicular bisector of BC. Where these two perpendicular bisectors intersect, mark D.

With the point of your compasses on D and with a radius equal to the length of AD, draw a circle. Does this circle pass through B and C?

_____

In questions 1 and 2 use a scale of 1 cm to 1 m to make a scale drawing. You should carry out these drawings outside of your workbook.

1.

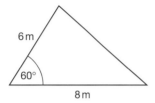

6 m
60°
8 m

2.

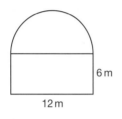

6 m
12 m

For questions 3 and 4 choose your own scale. This scale should give lines that are long enough to draw easily. Do not use a scale that leads to awkward fractions of a centimetre, such as thirds.

3.

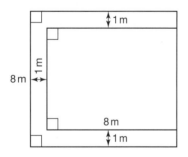

1 m
1 m
8 m
8 m
1 m

4.

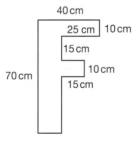

40 cm
25 cm   10 cm
15 cm
10 cm
70 cm
15 cm

In questions 5 and 6, A is a point on the ground, Â is the angle of elevation of C, the top of BC. Using a scale of 1 cm ≡ 5 m, make a scale drawing to find the height of BC.

5.

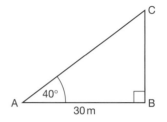

C
40°
A
30 m
B

6.

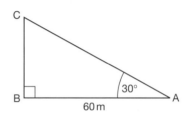

C
B
30°
A
60 m

7.

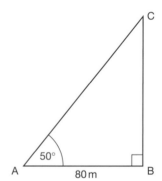

C
50°
A
80 m
B

From a point A on the ground the angle of elevation of a point C is 50°. Using a scale of 1 cm ≡ 10 m, make a scale drawing and hence find the height BC.

BC = _____

8. An office block is 64 m high. From a point P on the ground, the angle of elevation of the top of the block is 36°. Using a scale of 1 cm ≡ 10 m, make a scale drawing and hence find the distance of P from the foot of the block.

_____

9.

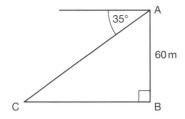

From A, the angle of depression of C is 35°. Use a scale of 1 cm ≡ 10 m, to make a scale drawing and hence find the distance BC.

BC = _____

10. From the top of a cliff which is 50 m high, the angle of depression of a yacht out at sea is 32°. Use your own scale to make a scale drawing. How far is the yacht from the base of the cliff?

Distance from foot of cliff = _____m

11. Viewed from the top of a cliff 45 metres high the angles of depression of two boats directly out to sea are 42° and 29°. Using 1 cm ≡ 5 m make a scale drawing and use it to find:

    a the distance of each boat from the base of the cliff

    _____

    b the distance between the boats. _____

In questions 12 to 16 draw a rough sketch to show each of the following bearings. Mark the angle in your sketch.

12. From a point A the bearing of my school, S, is 075°.

13. The bearing of the town hall, H, from a point B, is 240°.

14. From a point Q the bearing of the post office, P, is 155°.

15. The bearing of the harbour office, H, from a boat B, is 320°.

16. From a village, R, the bearing of a village, S, is 100°.

17. From a ship, S, a lighthouse, L, is 500 m on a bearing of 033°, and a trawler, T, is 750 m on a bearing of 123°. Using a scale of 1 cm ≡ 100 m, make a scale drawing and use it to find the distance and bearing of the lighthouse from the trawler.

Distance = _____m

Bearing = _____°

18.

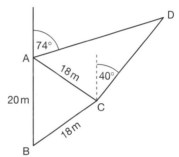

A and B show the positions of the wickets, which are 20 m apart, on a cricket pitch. The batsman at A strikes the ball towards D on a bearing of 074°.

A fielder, standing at C, which is 18 m from each wicket, runs at a bearing of 040°, to cut off the ball. The fielder intercepts the ball at D.

Use a scale of 1 cm ≡ 2 m to make a scale drawing and from it find:

    a the distance the fielder has to run before he retrieves the ball

    _____

    b the distance travelled by the ball along the ground from bat to fielder.

    _____

19.

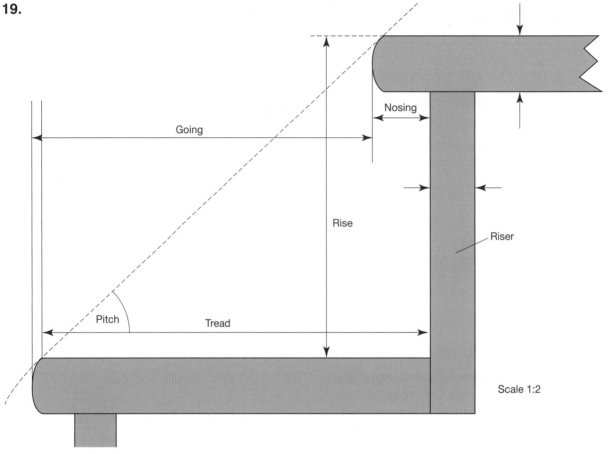

This is an accurate drawing, drawn half-size, of the cross-section of part of a domestic staircase.

a Use the drawing to find the actual length of:

   i the tread   _____

   ii the rise   _____

   iii the going   _____

   iv the nosing.   _____

b What thickness of timber has been used for:

   i the tread   _____

   ii the riser?   _____

c A safety regulation states that the sum of the going plus twice the rise must be more than 500 mm but less than 700 mm. Do these stairs satisfy this regulation?

   Give a reason for your answer.

   _____

d The maximum pitch of a staircase is 42°.

   Is this regulation satisfied? Give a reason for your answer.

   _____

In questions 1 to 20 circle the letter for the correct answer.

1. The value of $12 - 24 \div 4 + 7$ is

    A −5   **B** 4

    C 11   **D** 13

2. For the number 52.654 the third significant figure is

    A 2   **B** 4

    C 5   **D** 6

3. In standard form the number 5 670 000 is

    A 56.7×10^5 **B** 5.67×10^6

    C 5.6×10^6 **D** 5.67×10^5

4. Written as a decimal $\frac{5}{8}$ is

    A 0.6   **B** 0.62

    C 0.625   **D** 0.63

5. The correct inequality between the two values $\frac{1}{3}$ and 0.333 is

    A $>$ **B** $\geqslant$

    C $<$ **D** $\leqslant$

6. The LCM of 4, 5 and 6 is

    A 30   **B** 60

    C 120   **D** 180

7. What is the value of $(-12) \div (-4) + 3$?

    A −4   **B** 0

    C 6   **D** 12

8. The value of x that satisfies the equation $5x = 3x - (x - 9)$ is

    A −3   **B** 0

    C 1   **D** 3

9. If $x = 0.26$ correct to 2 d.p. then

    A $0.2 \leqslant x \leqslant 0.3$ **B** $0.255 \leqslant x < 0.265$

    C $0.255 \leqslant x \leqslant 0.265$ **D** $0.25 \leqslant x < 0.26$

10. $5(a - b) - 3(a - b)$ simplifies to

    A $2a - 2b$ **B** $2a + 2b$

    C $2a + 8b$ **D** $8a - 2b$

11. Given that $P = 3q - 2r$ when $q = \frac{1}{2}$ and $r = \frac{1}{4}$, $P =$

    A $\frac{1}{4}$ **B** $\frac{1}{2}$

    C 1   **D** 2

12. Two angles of an isosceles triangle are 54° and 63°. The size of the third angle is

    A 54°   **B** 60°

    C 63°   **D** 72°

13. Given that $I = \dfrac{PRT}{100}$ then T expressed in terms of the other letters is

    A $\dfrac{100P}{RI}$ **B** $\dfrac{PR}{100I}$

    C $\dfrac{100I}{PR}$ **D** $\dfrac{100R}{PI}$

14. 34% of 19.2 is

    A 12.672   **B** 6.72

    C 6.53   **D** 6.528

15. The selling price of an article sold at a loss of 35% is US $156. The cost price is

    A US $210.60   **B** US $240

    C US $257.40   **D** US $446

16. Which of the following pairs is the solution of the simultaneous equations

$4x + y = 10$

$2x + y = 4$

    A $(-2, 3)$ **B** $(0, 3)$

    C $(3, 0)$ **D** $(3, -2)$

17. When $250 000 was invested at 5% simple interest for a number of years it amounted to $325 000. The number of years for which it was invested was

    A 4   **B** 5

    C 6   **D** 7

18. If the bearing of Q from P is 225°, the bearing of P from Q is

    A 045°   **B** 135°

    C 225°   **D** 315°

19. 400 grams as a percentage of 2 kilograms is

    A 10%   **B** 15%

    C 20%   **D** 25%

20. $12\frac{1}{2}\%$ of $1120 is

    A $14   **B** $149

    C $280   **D** $140

21.

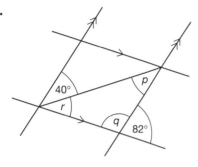

Find each marked angle.

$p =$ _____ $q =$ _____ $r =$ _____

22. Outside your workbook, construct a triangle ABC in which AB = 9 cm, AC = 7 cm and BÂC = 30°. Construct the perpendicular, CE, from C to AB. Measure and record its length.

CE = _____

23. a Make a the subject of the formula $v = u + at$.

_____

    b If $p = qr - s^2$ find p when $q = 5$, $r = 2$ and $s = 3$.

_____

24. a How many Jamaican dollars will an American tourist get for US$725 when the rate of exchange is US$1 to 80 Jamaican dollars?

_____

    b Change J$4500 into UK pounds if J$1 = £0.0082.

_____

25. a Last year Peter Morgan spent 63% of his income on household expenses, 18% on his car, and 12% on other expenses.

He saved $197 400. Work out his annual income.

_____

    b When $8\frac{1}{2}\%$ of the students in a school are absent, 1098 students are present. How many students are there in the school?

_____

26. a Arrange the numbers 0.83, $\frac{5}{6}$, $\frac{7}{9}$, 0.7 in ascending order.

_____

    b Arrange the numbers $\frac{8}{9}$, 0.88, 0.77, $\frac{7}{9}$ in descending order.

_____

27. Find:

    a $1\frac{7}{8} - 2\frac{1}{4} + 3\frac{3}{4}$ _____

    b $1\frac{5}{7} \times \frac{1}{2} \div 3\frac{3}{7}$ _____

28. Simplify:

    a $2^4 \times 2^7$ _____

    b $2^7 \div 2^4$ _____

    c $(2x^3y^2)^3$ _____

    d $(3a^2b^4)^2$ _____

29. a A motorcycle is bought for $360 000 and depreciates 10% a year. What is it worth at the end of its second year?

_____

    b A store gives a discount of 5% for a cash sale. I buy a TV marked $64 500 and pay cash. What must I pay for it?

_____

30. From a point 20 m from the foot of a vertical statue the angle of elevation of the top of the statue is 26°. Outside your workbook, show this information on a scale drawing. Use 1 cm to represent 2 m.

How high is the statue? _____

8 Straight Line Graphs

1.

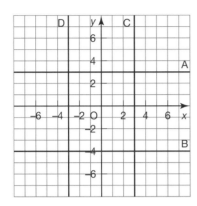

Give the equations of the straight lines marked A, B, C and D.

Equation of line A   _____

Equation of line B   _____

Equation of line C   _____

Equation of line D   _____

2. Draw on the diagram, the lines:

A with equation $x = -3$

B with equation $y = 6$

C with equation $x = 4$

D with equation $y = -2$

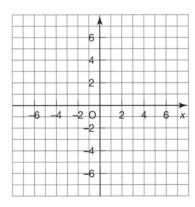

3. For each question complete the table. Then, draw, on graph paper, x and y axes for the range indicated in brackets. Mark three points and draw a straight line through them.

a $y = x + 3$ $(-4 \leqslant x \leqslant 4, -2 \leqslant y \leqslant 6)$

x	–3	0	3
y			

b $y = 3 - x$ $(-3 \leqslant x \leqslant 4, -2 \leqslant y \leqslant 6)$

x	–2	0	4
y			

4. For each question choose your own values within the given range. Construct a table and draw the line.

a $y = 4 - 2x$ $(-2 \leqslant x \leqslant 3, -2 \leqslant y \leqslant 6)$

x			
y			

b $y = \frac{1}{2}x + 2$ $(-4 \leqslant x \leqslant 4, 0 \leqslant y \leqslant 5)$

x			
y			

5. Draw the x and y axes using the given range for the x-axis and the range you choose for the y-axis. Draw the line:

a $y = 2x - 1$ **b** $y = 1 - 2x$
   $(-2 \leqslant x \leqslant 3)$ $(-2 \leqslant x \leqslant 2)$

6.

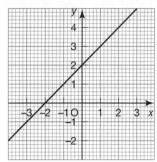

You are given the graph of the line $y = x + 2$.
From the graph find:

a y if $x = \frac{1}{2}$ _____

b x if $y = 1.2$ _____

c x if $y = -0.4$ _____

7.

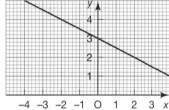

You are given the graph of the line $y = 3 - \frac{1}{2}x$.
From the graph find:

a y if $x = 1$ _____

b x if $y = 3.6$ _____

c y if $x = -2.5$ _____

8. Find whether the given points lie on the line
whose equation is given:

a $y = 1 + 2x$; (2, 5) _____ (3, 6) _____

b $y = 5 - 2x$; (4, 3) _____ (3, -1) _____

c $y = 8 - \frac{1}{2}x$; (4, 4) _____ (4, -4) _____

9. Draw, on the same pair of axes, the graphs of
the lines:

$y = \frac{1}{2}x - 2, y = \frac{1}{2}x + 2, y = \frac{1}{2}x + 4,$

for $-4 \leqslant x \leqslant 5, -5 \leqslant y \leqslant 7.$

What do you notice? _____

10. Draw on graph paper and on the same pair of
axes, the graphs of the lines:

$y = 5 - 2x, y = 7 - 2x, y + 2x = 0,$

for $-2 \leqslant x \leqslant 3, -6 \leqslant y \leqslant 11.$

What do you notice? _____

11. Find the gradients of the straight line through
each of the following pairs of points:

a (5. 2), (7, 6) _____

b (3, 1), (1, 7) _____

c (2, 3), (7, -5) _____

d (-2, 5), (-5, 3) _____

12. If lines are drawn through the following points,
which lines have zero gradient and which are
parallel to the y-axis?

a (-3, 5), (5, 5) _____

b (4, -2), (4, 6) _____

c (-2, -3), (-5, -3) _____

13. Sketch a line with gradient:

a 2

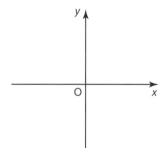

b -2

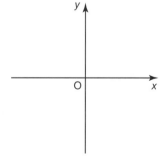

c $\frac{1}{2}$

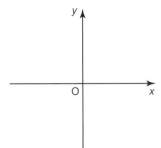

d $-\frac{1}{2}$

14. Give the gradients and *y*-intercept of the lines with the following equations:

a $y = 4x - 7$ _____

b $y = 3 - 4x$ _____

c $y = 10 + \frac{1}{3}x$ _____

d $y = 2x - 3$ _____

e $y = 8 - \frac{2}{3}x$ _____

15. Which of the given lines are parallel?

$y = 4 - 2x$, $y = 2x + 3$, $y = 5 - 2x$,
$2y = 4x + 7$, $y = 5 + 2x$.

_____

16. What is the gradient of the line $y = 4 - 3x$?

_____

Find the equation of the straight line that is parallel to the given line and passes through the point:

a (0, 0) _____

b (0, 2) _____

c (2,0) _____

d (2, 2) _____

17. A line of gradient $-\frac{1}{2}$ passes through the origin.

a Find its equation.

_____

b Find the equation of the line that is parallel to this line and passes through the point (0, −4).

_____

18. On graph paper draw the lines with the following equations. Use 1 cm ≡ 1 unit. For each line write down its gradient.

a $3x + 4y = 12$ $0 \leqslant x \leqslant 6$

_____

b $x - 3y = 9$ $0 \leqslant x \leqslant 10$

_____

c $2x + 5y = 10$ $0 \leqslant x \leqslant 6$

_____

d $4x - 2y = 15$ $0 \leqslant x \leqslant 5$

_____

19. Draw on graph paper the lines whose equations are given below. Write down the gradient of each line.

a $\frac{x}{2} + \frac{y}{5} = 1$ _____

b $\frac{x}{3} + \frac{y}{6} = 1$ _____

c $\frac{x}{4} + \frac{y}{5} = 1$ _____

20. Find the gradient and intercept on the *y*-axis of each of the following lines:

a $x - 3y = 7$ _____

b $5x + 3y = 15$ _____

c $2x + 3y = 15$ _____

d $4x + 7y = 28$ _____

21. Find the gradient and intercept on the y-axis of each of the following lines:

   a $\dfrac{x}{5} + \dfrac{y}{2} = 1$ _____

   b $\dfrac{x}{7} + \dfrac{y}{9} = 1$ _____

   c $\dfrac{x}{4} - \dfrac{y}{5} = 1$ _____

   d $x + y = 4$ _____

   e $x + y = -5$ _____

   f $3x + 5y = 15$ _____

22. Find the gradient and intercept on the y-axis of the line through the given points. Hence find the equation of the line.

   a   (0, 2) and (3, 0)

   _____

   b   (1, 0) and (4, 3)

   _____

   c   (0, –4) and (1, 5)

   _____

   d   (0, 8) and (4, 0)

   _____

23. Find the gradient, and equation, of the straight line through the given points:

   a   (3, 1) and (6, 10)

   _____

   b   (2, –4) and (4, –2)

   _____

   c   (–1, 5) and (1, 9)

   _____

   d   (–2, 3) and (2, –3)

   _____

24. On the same pair of axes, for $-1 \leqslant x \leqslant 2$ and $-2 \leqslant y \leqslant 3$, draw the graphs of the lines

$y = 3x, \quad y = 3x - 2, \quad y = 3 - 3x, \quad y = -3x.$

(Use $4\,\text{cm} \equiv 1$ unit on both axes.)

Name the type of quadrilateral bounded by these lines.

_____

9 Graphs

In this unit, use graph paper to draw any supporting graphs and tables.

1. The table shows the capacity (C litres), for containers that are mathematically similar, but have different depths (d cm).

Depth (d cm)	10	12	15
Capacity (C litres)	0.3	0.52	1

Depth (d cm)	17	18	20
Capacity (C litres)	1.47	1.75	2.4

Draw a graph connecting d and C using 4 cm ≡ 5 units on the horizontal d-axis and 4 cm ≡ 1 unit on the C-axis. Use your graph to find:

a the depth of a similar container with a capacity of 2 litres

_____

b the capacity of a similar container that is 13.5 cm deep.

_____

2. The velocities, V metres per sec, of waves at varying depths of water, d metres, are given in the table.

d (m)	1	2	3	4
V (m/s)	3.2	4.5	5.6	6.6

d (m)	6	8	10
V (m/s)	8.1	9.3	10.2

Draw a graph connecting d and V.

(Use 2 cm ≡ 1 unit on both axes.) Let 2 be the lowest value on the V-axis. Use your graph to find:

a the velocity of a wave that is:

    i 4.5 metres deep _____

    ii 9 metres deep. _____

b the depth of a wave with a velocity of:

    i 5 metres per second _____

    ii 8.4 metres per second. _____

3. The table shows the torque, or turning power, T newton metres, at different speeds, R revolutions per minute, for a new engine.

Speed (R r.p.m.)	1500	2000	3000
Torque (T n.m.)	115	143	155

Speed (R r.p.m.)	4000	5000	6000
Torque (T n.m.)	145	128	108

Draw a graph connecting R and T. (Use 4 cm ≡ 1000 units on the R-axis and 4 cm ≡ 50 units on the T-axis.) Let 1000 be the lowest value on the R-axis and 50 the lowest value on the T-axis. Use your graph to find:

a the maximum value of T and the value of R for which it occurs

_____

b the values of R when $T = 130$.

_____

4.

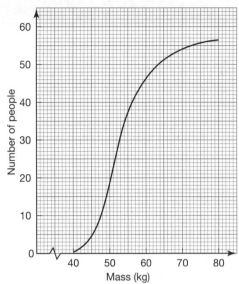

Mass (kg)

This graph shows the masses of a group of people on an outing. It shows the number of people up to a given mass. For example the mass of 5 people was less than 45 kg.

a How many people gave their masses?

_____

b How many people are a mass less than 60 kg?

_____

c Find the number of people whose mass is 55 kg or more.

_____

5. Two quantities X and Y are connected by the formula $Y = \dfrac{20}{X}$. Construct a table to show the values of Y for values of X from 1 to 10. Plot these points on a graph and from this graph find:

a the value of Y when $X = 3.5$

_____

b the value of X when $Y = 12$.

_____

6. Draw the graph of $y = x^2 - 4$ for values of x in the range −3 to 3. Take 4 cm as the unit for x and 2 cm as the unit for y.

a Use your graph to find the values of x when:

i $y = -3$ _____

ii $y = 4$ _____

b Are there any values of x for which $y = 0$?

7. Draw the graph of $y = 2x(x - 5)$ for values of x in the range −1 to 6. Make a table for values of y taking values of x at unit intervals. Use a scale of 2 cm for 1 unit on the x-axis and 1 cm for 1 unit on the y-axis. Use your graph to find:

a the values of x where the graph crosses the x-axis

_____

b the lowest value of $2x(x - 5)$ and the corresponding value of x.

_____

8. Draw the graph of $y = 6 - 3x - x^2$ for values of x in the range −5 to 2. Make a table for values of y taking values of x at unit intervals. Take 2 cm as 1 unit for x and 1 cm as 1 unit for y. Use your graph to find:

a the highest value of $6 - 3x - x^2$ and the corresponding value of x

_____

b the value of x when $6 - 3x - x^2$ has a value of

i 7 _____

ii −3 _____

10 Areas and Volumes

1. Find the area of this triangle.

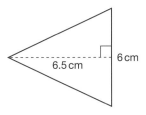

2. Find the area of this parallelogram.

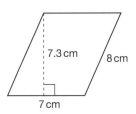

In questions 3 to 5 use squared paper. Draw axes for *x* and *y* in the ranges $-6 \leqslant x \leqslant 6$, $-6 \leqslant y \leqslant 6$.

Use 1 square for 1 unit. Draw the figure and find its area in square units:

3. Rectangle ABCD with A(−2, 4), B(6, 4) and C(6, −2).

Write down the coordinates of D.

4. Square ABCD with A(−5, 3), B(3, 3) and C(3, −5).

Write down the coordinates of D.

5. Triangle ABC with A(−4, −4), B(5, −4) and C(2, 5).

In questions 6 and 7 give answers correct to 3 s.f. where necessary.

6.

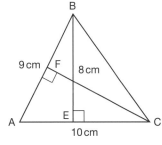

In △ABC, AB = 9 cm, AC = 10 cm and BE = 8 cm. Find:

a the area of △ABC

b the length of CF.

7.

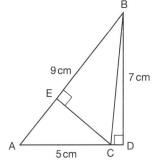

In △ABC, AB = 9 cm, AC = 5 cm and BD = 7 cm. E is the foot of the perpendicular from C to AB. Find:

a the area of △ABC

b the length of CE.

8.

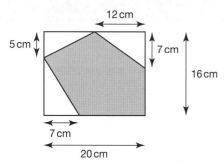

Three triangular pieces are removed from a rectangle measuring 20 cm by 16 cm.

Find the area remaining.

It is shown shaded.

_____

9.

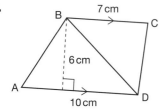

ABCD is a trapezium. Find its area.

_____

10.

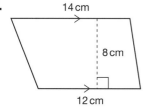

Find the area of this trapezium.

_____

11.

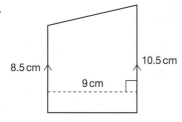

Find the area of this trapezium.

_____

In questions 12 to 14 use squared paper. Draw axes for x and y in the ranges $-6 \leqslant x \leqslant 6$, $-6 \leqslant y \leqslant 6$. Use 1 square for 1 unit. Plot the points and join them in alphabetical order. Find, in square units, the area of the resulting shape:

12. A(3, –2), B(6, 2), C(–4, 2), D(2, –2)

_____

13. A(3, 6), B(–4, 5), C(–4, –1), D(3, –4)

_____

14. A(5, –4), B(5, 1), C(1, 5), D(–6, 3), E(–6, –4).

_____

15. Using squared paper draw x and y axes for

$-6 \leqslant x \leqslant 8$, $-6 \leqslant y \leqslant 8$.

Draw parallelogram ABCD where A(1, 1), B(1, 7), C(7, 4), D(7, –2).

Using AB as the base in each case draw three other parallelograms whose areas are equal to the area of ABCD.

16.

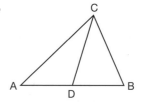

D is the midpoint of AB.

Show that area of △ACD = area of △DBC.

_____

_____

17.

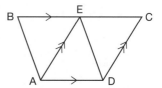

E is the midpoint of BC.

Show that area of △ABE = area of △ECD.

_____

_____

18.

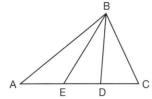

D and E are points on the base BC of a triangle ABC such that AE = ED = DC. Find the ratio of

a area of △ABE : area of △ABC

_____

b area of △BCE : area of △ABC

_____

19.

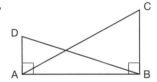

The area of triangle ABD is $\frac{5}{12}$ the area of triangle ABC. Given that AD = 35 cm find the length of BC.

_____

20.

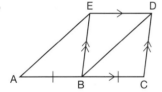

ABDE and BCDE are parallelograms.

Show that the area of △ABE = area of △BCD

_____

21.

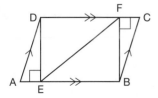

ABCD is a parallelogram. EBFD is a rectangle. Show that:

a area of △ADE = area of △BCF

_____

b area of △AEFD = area of △EBCF.

_____

22. On graph paper, construct a parallelogram ABCD with AB = 5 cm, BC = 7 cm and $A\hat{B}C = 60°$.

Now, construct a parallelogram ABEF that is equal in area to ABCD such that BE = 8 cm and E and D are on opposite sides of BC.

Measure $A\hat{B}E$. _____

23. For each sector find:

i the length of the arc subtended by the given angle

ii the area of the sector.

a

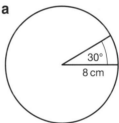

b

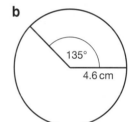

i _____   **i** _____

ii _____   **ii** _____

c

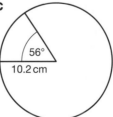

d

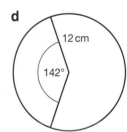

i _____   **i** _____

ii _____   **ii** _____

31

24.

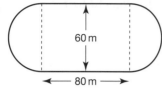

A field used for games is laid out as a rectangle measuring 80 m by 60 m with semi–circles at both ends. Find:

a its perimeter _____

b its area. _____

25.

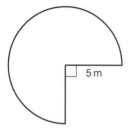

A flower–bed is in the shape of a circle of radius 5 m from which a quadrant has been removed. Find:

a its perimeter _____

b its area. _____

26.

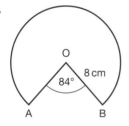

A sector is removed from a circular sheet of metal of radius 8 cm. If AOB = 84° find:

a the perimeter of what remains

_____

b its area. _____

27.

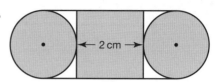

The diagram shows 2 pieces of wooden dowel, each of diameter 2 cm. They are held against a third piece of wood which is square in cross–section, by an elastic band. The stretched length of the elastic band is 50% more than its unstretched length. Find:

a its stretched length _____

b its unstretched length. _____

28. Convert the given quantity into the unit given in brackets.

a $42\,cm^2$ (mm^2) _____

b 60 ml ($c\ell$) _____

c $0.006\,cm^3$ (mm^3) _____

d 0.07 litres ($c\ell$) _____

e $80\,000\,cm^3$ (m^3) _____

29. a Find the volume of a cuboid measuring 10 cm by 8 cm by 3 cm.

_____

b Find the volume of a cuboid measuring 150 cm by 30 cm by 30 cm.

_____

30. Find the volumes of the following cuboids, changing the unit first if necessary. Do not draw a diagram.

The units for your answers are given.

	Length	Width	Height	Volume
a	25 cm	6 cm	4 cm	cm^3
b	8.2 cm	0.5 cm	20 mm	cm^3
c	30 mm	10 mm	7 mm	m^3
d	2.6 m	0.45 m	25 cm	m^3
e	7.5 m	3.6 m	1.6 m	m^3

Find the volumes of the following prisms. Draw a diagram of the cross–section but do not draw a picture of the solid.

31.

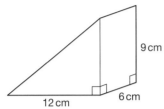

9 cm

12 cm   6 cm

32.

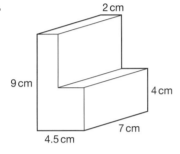

2 cm

9 cm   4 cm

7 cm

4.5 cm

33.

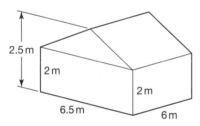

2.5 m

2 m

6.5 m   2 m

6 m

34. The cross–section and length of the prism is given. Find its volume.

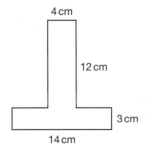

4 cm

12 cm

3 cm

14 cm

Length 12 cm

Volume _____

In the following questions use the value of π on your calculator. Give all your answers correct to 3 s.f.

35. Find the volumes of the following cylinders:

a Radius 5 cm, height 6 cm

_____

b Radius 8 cm, height 5.5 cm

_____

c Radius 3.5 cm, height 8.5 cm

_____

36. Find the volumes of the following cylinders:

a Diameter 12 cm, height 3.6 cm

_____

b Diameter 3.5 cm, height 3 cm

_____

c Diameter 0.6 m, height 0.25 m

_____

37.

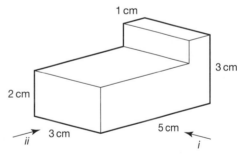

1 cm

3 cm

2 cm

ii   3 cm   5 cm   *i*

This solid is built up of cubes with sides of 1 cm.

Draw, on squared paper, the plan and the elevations in the directions indicated by the arrows.

1. Find the arithmetic average or mean of the following sets of numbers:

   a   56, 74, 92, 63, 70

   _____

   b   3.7, 5.8, 3.4, 8.2, 7.5, 6.9, 5.8

   _____

2. The heights of five adults are 179 cm, 183 cm, 177 cm, 193 cm, 153 and 171 cm.

   a   Find the total of these heights.

   _____

   b   Find the average height.

   _____

3. The mid-day temperatures at a resort during a week in January were 34°, 22°, 32°, 24°, 32°, 29°, 30°. What was the average mid-day temperature at the resort for the week?

   _____

4. The Saturday takings of a small shop over a four-week period were: $212 000, $164 000, $178 000 and $174 800. Find the average Saturday takings over the four weeks.

   _____

5. In a dancing competition the marks awarded to a couple were: 5.4, 4.9, 5.8, 5.5, 5.7, 5.5, 5.5, 5.4. Their score is found by averaging these marks after discounting the highest mark and the lowest mark. Calculate the couple's score.

   _____

6. The average age of five students is 19 years 7 months. The average age of four of them is 18 years 11 months. How old is the fifth student?

   _____

7. The average daily rainfall for a week in Totshill was 4.6 mm. For the first six days of that week the average daily rainfall was 1.5 mm. How much rain fell on the seventh day?

   _____

8. The average mark in English for a class of 26 students was 73%. The average mark in English for another class of 32 students was 65%. What was the average mark for the combined classes?

   _____

9. In the first 20 completed innings of a season a batsman scored 1124 runs. In the next innings he was out for 136. By how much did this increase his average?

   _____

10. A Test bowler takes 110 wickets for 1267 runs. Find his bowling average correct to 2 decimal places.

    _____

11. Last season a batsman scored 976 runs in 38 completed innings. This season he has scored 1242 in 43 completed innings. Calculate his average over the two seasons giving your answer correct to 1 decimal place.

    _____

12. Find the average speed in km/h for a journey of:

    a   45 km in $1\frac{1}{2}$ hours   _____

    b   60 km in 45 minutes   _____

    c   80 km in 1 hours 20 minutes.   _____

13. How far will a vehicle travel

   a in 3 hours at an average speed of 34 m.p.h.

   b in $1\frac{1}{2}$ hours at an average speed of 64 km/h

   c in 45 minutes at an average speed of 192 km/h

   d in 10 minutes at an average speed of 120 m.p.h.?

14. How long will it take to travel

   a 88 miles at an average speed of 33 m.p.h.

   b 57 km at an average speed of 18 km/h

   c 49 km at an average speed of 84 km/h?

15. George drives for 2 hours at an average speed of 60 km/h, and then for 45 minutes at an average speed of 16 km/h. Find his average speed for the whole journey.

16. A coach travels from Walltown to Portborough, a distance of 10 miles, at an average speed of 20 m.p.h., and continues its journey to Riverton, a further 20 miles beyond Portborough, at an average speed of 24 m.p.h. Find its average speed for the whole journey from Walltown to Riverton.

17. A motorist wanted to make a journey of 110 kilometres in 2 hours. He travelled the first 60 kilometres at an average speed of 45 km/h, and the next 30 kilometres at an average speed of 90 km/h. Work out his average speed for the remaining 20 kilometres if he is to arrive on time.

12 Algebraic Products

1. Expand:

a $4(x + 3)$ _____

b $5(x - 4)$ _____

c $7(a + 2)$ _____

d $3(b + 5)$ _____

e $2(3a + 1)$ _____

f $6(2a + 3)$ _____

2. Expand:

a $5x(y + 2z)$ _____

b $3a(2b + 3c)$ _____

c $4x(2y - 5z)$ _____

d $7a(b - 4c)$ _____

e $2x(3y + 5z)$ _____

f $8x(y - 3z)$ _____

3. Expand:

a $(a + b)(c - d)$ _____

b $(p + q)(r + s)$ _____

c $(x - y)(z + 3)$ _____

d $(x + y)(z - 5)$ _____

e $(a - b)(c - d)$ _____

f $(p - q)(r - s)$ _____

4. Expand:

a $(p + q)(2r - 3s)$ _____

b $(a + 2b)(2 - z)$ _____

c $(3x - 2y)(2z + 1)$ _____

d $(x - y)(z + 5)$ _____

e $(7a - 2b)(2c + 3d)$ _____

f $(p + q)(2r - 3s)$ _____

5. Expand:

a $(x + 1)(x + 7)$ _____

b $(x + 3)(x + 5)$ _____

c $(x + 10)(x + 1)$ _____

d $(a + 5)(a + 7)$ _____

e $(b + 2)(b + 5)$ _____

6. Expand:

a $(x - 1)(x - 4)$ _____

b $(x - 5)(x - 3)$ _____

c $(x - 7)(x - 4)$ _____

d $(a - 6)(a - 2)$ _____

e $(b - 8)(b - 2)$ _____

7. Expand:

a $(x + 2)(x - 4)$ _____

b $(x - 3)(x + 5)$ _____

c $(x + 7)(x - 1)$ _____

d $(a - 9)(a + 3)$ _____

e $(b + 4)(b - 6)$ _____

8. Expand:

a $(x + 2)(x - 6)$ _____

b $(a + 9)(a + 7)$ _____

c $(c - 3)(c - 4)$ _____

d $(y - 1)(y + 3)$ _____

e $(t + 4)(t - 5)$ _____

9. Expand the following products:

a $(2a + 1)(a + 2)$ _____

b $(3x + 2)(x + 1)$ _____

c $(5x + 3)(x + 2)$ _____

d $(x + 2)(3x + 3)$ _____

e $(x + 4)(5x + 3)$ _____

10. Expand:

a $(4x + 3)(2x + 1)$ _____

b $(2x + 4)(3x - 2)$ _____

c $(3a + 4)(5a - 3)$ _____

d $(5x - 3)(4x - 7)$ _____

e $(2s - 5)(3s + 9)$ _____

11. Expand:

a $(2x + 1)(3 + 4x)$ _____

b $(4x - 3)(4 - 3x)$ _____

c $(5x + 1)(2 - 5x)$ _____

d $(7x - 3)(5 + 2x)$ _____

e $(3x + 5)(7 - x)$ _____

12. Expand:

a $(x + 5)^2$ _____

b $(a + 2)^2$ _____

c $(p + 3)^2$ _____

d $(s + t)^2$ _____

e $(y + z)^2$ _____

13. Expand:

a $(2a + 1)^2$ _____

b $(3x + 2)^2$ _____

c $(2x + 7)^2$ _____

d $(5x + 2y)^2$ _____

e $(p + 5q)^2$ _____

14. Expand:

a $(x - 4)^2$ _____

b $(a - 7)^2$ _____

c $(y - z)^2$ _____

d $(b - c)^2$ _____

e $(s - t)^2$ _____

15. Expand:

a $(4x - 1)^2$ _____

b $(7a - 3)^2$ _____

c $(5y - 1)^2$ _____

d $(3x - 2)^2$ _____

e $(8x - 1)^2$ _____

16. Expand:

a $(x - 2y)^2$ _____

b $(2a - 3b)^2$ _____

c $(5x - 2y)^2$ _____

d $(3a - 4b)^2$ _____

e $(2p - 7q)^2$ _____

17. Expand:

    a $(x - 3)(x + 3)$ _____

    b $(a + 5)(a - 5)$ _____

    c $(b - 7)(b + 7)$ _____

    d $(5y + 4)(5y - 4)$ _____

    e $(3x + 1)(3x - 1)$ _____

18. Expand:

    a $(2x - 3y)(2x + 3y)$ _____

    b $(5a + 2b)(5a - 2b)$ _____

    c $(4y - 3z)(4y - 3z)$ _____

    d $(7a + 5b)(7a - 5b)$ _____

    e $(8x - 5y)(8x + 5y)$ _____

19. Simplify:

    a $(x + 1)(x + 2) + x(x + 3)$

    _____

    b $x(x + 5) + (x + 2)(x + 3)$

    _____

    c $(a - 3)(a - 5) + a(a + 2)$

    _____

    d $(x + 4)(x - 3) - x(x - 7)$

    _____

    e $(a - 7)(a - 5) - a(a - 9)$

    _____

20. Expand:

    a $(pq - 4)^2$ _____

    b $(2ab + 3)^2$ _____

    c $(xy + 7)^2$ _____

    d $(xy - z)^2$ _____

    e $(4 - 3yz)^2$ _____

13 Algebraic Factors

1. Factorise:

   a $5x + 5$ _____

   b $4a - 8$ _____

   c $7x - 14$ _____

   d $9b + 27$ _____

2. Factorise:

   a $x^2 + 5x$ _____

   b $x^2 - 9x$ _____

   c $a^2 + 7a$ _____

   d $3x^2 + x$ _____

   e $5a^2 - a$ _____

3. Factorise:

   a $4x^2 + 2x$ _____

   b $10ab - 5bc$ _____

   c $27y^2 + 9y$ _____

   d $3x^2 - 9x$ _____

   e $12p^2 - 4p$ _____

4. Factorise:

   a $3x^2 + 6x + 9$ _____

   b $4x^2 + 8x - 12$ _____

   c $3x - 6y + 9z$ _____

   d $25x^2 + 10x + 5$ _____

   e $pq + 3qr - 3qs$ _____

5. Factorise:

   a $a^3 + a^2$ _____

   b $5x^3 + 10x$ _____

   c $15x^2 - 25x^4$ _____

   d $16 + 4y^2$ _____

   e $9pq - 3qr$ _____

6. Factorise:

   a $4\pi r^2 + 2\pi rh$ _____

   b $\frac{1}{2}mv_1^2 + \frac{1}{2}mv_2^2$ _____

   c $mga - mgb$ _____

   d $\frac{1}{2}aH + \frac{1}{2}ah$ _____

   e $\frac{2}{3}\pi r^3 - \pi r^2 h$ _____

7. Factorise:

   a $xy - 2x - 4y + 8$ _____

   b $xy + 4x + y + 4$ _____

   c $ac - 5a + bc - 5b$ _____

   d $pr - qr + ps - qs$ _____

   e $15 - 5a + 3b - ab$ _____

8. Factorise:

   a $8a - 2ab - 4b + b^2$ _____

   b $16 - 2y - 8x + xy$ _____

   c $6a - 2ab - 9b + 3b^2$ _____

   d $2x - xy + 2y - y^2$ _____

   e $x^2 + xz + xy + yz$ _____

9. Factorise:

   a $2a + 3b - ab - 6$ _____

   b $xy - 3y - 4x + 12$ _____

c $a^2 + ab + 2a + 2b$ _____

d $2x^2 - 4xy + x - 2y$ _____

e $2x^2 + 4xy - x - 2y$ _____

10. Factorise:

a $x^2 + 10x + 24$ _____

b $x^2 + 10x + 9$ _____

c $x^2 + 10x + 16$ _____

d $x^2 + 13x + 42$ _____

e $x^2 + 11x + 24$ _____

11. Factorise:

a $x^2 - 9x + 14$ _____

b $x^2 - 11x + 24$ _____

c $x^2 - 7x + 6$ _____

d $x^2 - 9x + 20$ _____

e $x^2 - 8x + 15$ _____

12. Factorise:

a $x^2 - 5x - 24$ _____

b $x^2 + 4x - 12$ _____

c $x^2 + 6x - 7$ _____

d $x^2 + x - 12$ _____

e $x^2 + 4x - 5$ _____

13. Factorise:

a $x^2 + 10x + 21$ _____

b $x^2 - 8x + 15$ _____

c $x^2 - 2x - 48$ _____

d $x^2 - 2x - 8$ _____

e $x^2 + 4x - 5$ _____

14. Factorise:

a $14 + x^2 - 9x$ _____

b $x^2 - 9 - 8x$ _____

c $30 - x - x^2$ _____

d $8x + x^2 - 48$ _____

e $11x - 26 + x^2$ _____

15. Factorise:

a $x^2 + 8x + 16$ _____

b $x^2 - 8x + 16$ _____

c $x^2 - 6x + 9$ _____

d $x^2 + 6x + 9$ _____

e $x^2 - 20x + 100$ _____

16. Factorise:

a $15 - 2x - x^2$ _____

b $42 + x - x^2$ _____

c $10 - 3x - x^2$ _____

d $2 - x - x^2$ _____

e $32 + 4x - x^2$ _____

17. Factorise:

a $x^2 - 9$ _____

b $9 - x^2$ _____

c $x^2 - 49$ _____

d $49 - x^2$ _____

e $16a^2 - b^2$ _____

18. Factorise:

a $4x + 12$ _____

b $6x - 14$ _____

c $6x^2 + 9$ _____

d $3x^2 + 18x + 15$ _____

e $4x^2 - 24x + 20$ _____

19. Factorise:

a $3x^2 + 2x - 1$ _____

b $4x^2 - 7x + 3$ _____

c $5x^2 - 13x + 6$ _____

d $3x^2 - 7x - 20$ _____

e $7x^2 + 3x - 4$ _____

20. Factorise:

a $6x^2 + 13x + 6$ _____

b $35x^2 + 34x + 8$ _____

c $21x^2 - 10x + 1$ _____

d $30x^2 + 29x + 7$ _____

e $5x^2 - 8xy - 13y^2$ _____

21. Factorise:

a $16x^2 - 1$ _____

b $25 - 9y^2$ _____

c $9x^2 - 16$ _____

d $25x^2 - 4y^2$ _____

e $100a^2 - 25b^2$ _____

22. Factorise:

a $5a^2 - 20b^2$ _____

b $6x^2 - 24$ _____

c $75x^2 - 27$ _____

d $\dfrac{x^2}{9} - \dfrac{y^2}{16}$ _____

e $\dfrac{25x^2}{9} - 1$ _____

23. Find without using a calculator:

a $3.5^2 + 0.5 \times 3.5$

b $5.27^2 - 4.73^2$

c $1.1 \times 5.9 + 5.9^2$

d $7.48^2 - 0.48 \times 7.48$

e $14.7^2 - 5.3^2$

24. Factorise where possible:

a $20x^2 - 9x + 1$ _____

b $6x^2 - 13x + 6$ _____

c $x^2 - 9x + 1$ _____

d $24x^2 + 44x + 12$ _____

e $6x^2 - x - 1$ _____

1.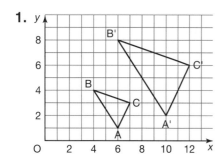

Draw AA', BB' and CC' and continue all three lines until they meet.

Write down the coordinates of the centre of enlargement.

_____

2.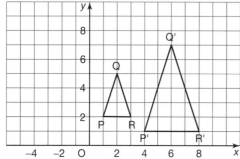

Draw PP', QQ' and RR' and continue all three lines until they meet.

Write down the coordinates of the centre of enlargement.

_____

3. Draw axes for x and y from 0 to 10.

Draw △ABC: A(3, 2), B(5, 0), C(6, 3).

Draw △A'B'C': A'(3, 4), B'(7, 0), C'(9, 6).

Draw AA', BB' and CC' and extend these lines until they meet.

a Give the coordinates of the centre of enlargement.

_____

b Measure the sides and angles of the two triangles. What do you notice?

_____

4. Draw axes for x from 0 to 8 and for y from 0 to 11.

Draw △ABC: A(2, 2), B(4, 4), C(5, 3).

Find the image, △A'B'C', of △ABC under an enlargement with centre (1, 3) and scale factor 2.

Write down the coordinates of △A'B'C'.

A' _____

B' _____

C' _____

5.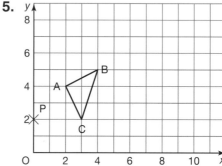

If P is the centre of enlargement draw the image, △A'B'C', of △ABC under an enlargement scale factor 2.

6. a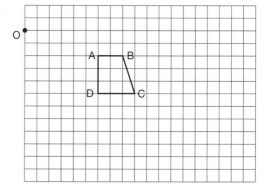

Using O as the centre of enlargement and a scale factor of 2, draw the enlargement of ABCD. Label the vertices A' B' C' D'. Check that the enlarged shape has sides twice as long as the original.

b

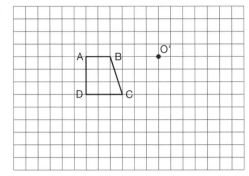

Repeat part (a) but this time use O' as the centre of enlargement. Is it true that the enlarged shape has sides twice as long as the original?

_____

What is the effect on the image of changing the centre of enlargement?

_____

7.

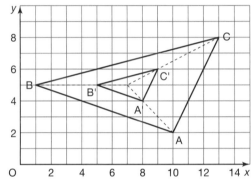

△A'B'C' is the image of △ABC.

Give the centre of enlargement and the scale factor.

Centre of enlargement _____

Scale factor _____

8. Draw axes for x and y from 0 to 9. Draw △ABC with A(0, 3), B(8, 3), C(8, 7) and △A'B'C' with A'(0, 3), B'(4, 3) and C'(4, 5).

Give the centre of enlargement and the scale factor.

Centre of enlargement _____

Scale factor _____

9. Draw axes for x and y from −2 to 11. Draw △ABC with A(8, 1), B(10, 5), C(6, 7).

Draw the image, A'B'C', of △ABC, with the point (−2, 0) as the centre of enlargement, and with a scale factor of $\frac{1}{2}$.

Write down the coordinates of △A'B'C'.

A' _____   B' _____   C' _____

10.

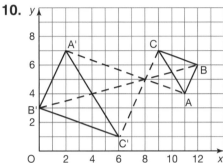

△A'B'C' is the image of △ABC.

Give the centre of enlargement and the scale factor.

Centre of enlargement _____

Scale factor _____

11. Draw x and y axes from −6 to 6. Draw △ABC with A(5, −2), B(3, −4), C(3, −2), and △A'B'C' with A'(−4, 1), B'(0, 5) and C'(0, 1).

Give the centre of enlargement and the scale factor.

Centre of enlargement _____

Scale factor _____

12. Draw axes for x and y from −3 to 11 using 1 cm as 1 unit on both axes.

Draw △ABC with A(8, 2), B(10, 6), C(6, 8). Find the image of △ABC under an enlargement, centre (−2, 2), scale factor $\frac{1}{2}$.

13. Draw axes for x and y from 0 to 12 using 1 cm as 1 unit on both axes.

Draw △ABC with A(10, 9), B(8, 10), C(8, 12). Find the image of △ABC under an enlargement, centre (7, 8), scale factor −2.

43

In questions **1** to **17** circle the letter for the correct answer.

1. Which of these points lie on the line
   $4x - 3y = 10$?

   i $(-1, 2)$
   ii $(1, -2)$
   iii $(4, 2)$

   A i and ii only   **B** i and iii only
   C ii and iii only   **D** i, ii and iii

2. What is the gradient of the line with equation
   $5x - 2y = 7$?

   A $-\frac{5}{2}$ **B** $-\frac{2}{5}$

   C $\frac{2}{5}$ **D** $\frac{5}{2}$

3. What is the gradient of the straight line joining
   the points $(-3, 2)$ and $(4, 1)$?

   A -1 **B** $-\frac{1}{7}$
   C $\frac{1}{7}$ **D** 1

4. For the line with equation $3x + 4y - 12 = 0$ the
   intercept on the y–axis is

   A -4 **B** -3
   C 3 **D** 4

5. The equation of the line passing through the
   points $(-1, 4)$ and $(4, -1)$ is

   A $y = \frac{5}{3}x + \frac{17}{3}$ **B** $y = \frac{3}{5}x + \frac{17}{5}$

   C $y = x + 3$ **D** $y = 3 - x$

6. $50\,000\,\text{cm}^3 =$
   A $5\,\text{m}^3$ **B** $0.5\,\text{m}^3$
   C $0.05\,\text{m}^3$ **D** $0.005\,\text{m}^3$

7.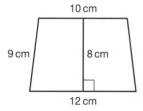

   The area of this trapezium is
   A $80\,\text{cm}^2$ **B** $88\,\text{cm}^2$
   C $99\,\text{cm}^2$ **D** $176\,\text{cm}^2$

8.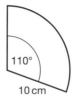

   The perimeter of this shape, correct to 1 d.p., is
   A $29.1\,\text{cm}$ **B** $29.2\,\text{cm}$
   C $39.1\,\text{cm}$ **D** $19.2\,\text{cm}$

9. Emma runs for 1 hour 40 minutes at an
   average speed of 12 km/h. The distance she
   runs is
   A $14\,\text{km}$ **B** $16\,\text{km}$
   C $18\,\text{km}$ **D** $20\,\text{km}$

10. A car travels 30 km at an average speed of
    40 km/h, then 90 km at an average speed of
    30 km/h. The average speed for the whole
    journey is
    A $64\,\text{km/h}$ **B** $60\,\text{km/h}$
    C $45\,\text{km/h}$ **D** $32\,\text{km/h}$

11. $(3a + 4b)(c - 5d) =$

    A $3ac + 15ad - 4bc - 20bd$
    B $3ac - 15ad + 4bc + 20bd$
    C $3ac + 15ad - 4bc + 20bd$
    D $3ac - 15ad + 4bc - 20bd$

12. $(x + 3)(2x - 5) =$

    A $2x^2 + x - 15$ **B** $2x^2 - x - 15$
    C $2x^2 + x + 15$ **D** $2x^2 - 11x - 15$

13. $(x - y)(x + 2z) =$

    A $x^2 - xy + 2xz - 2yz$
    B $x^2 - xy - 2xz + 2yz$
    C $x^2 - xy + 2xz + 2yz$
    D $x^2 - xy - 2xz - 2yz$

14. $(3a - 2)^2 =$

    A $6a^2 - 6a - 4$ **B** $6a^2 - 12a + 4$
    C $9a^2 - 12a + 4$ **D** $9a^2 - 6a + 4$

15. $x^2 + x - 20 =$

    A $(x + 10)(x - 2)$ **B** $(x + 4)(x - 5)$

    C $(x - 4)(x + 5)$ **D** $(x - 4)(x - 5)$

16. $15 - 2x - x^2 =$

    A $(3 + x)(5 - x)$ **B** $(3 - x)(5 + x)$

    C $(3 + x)(5 + x)$ **D** $(10 + x)(5 - x)$

17. $6x^2 + 13x - 28 =$

    A $(6x - 7)(x + 4)$ **B** $(2x - 7)(3x + 4)$

    C $(2x - 7)(3x - 4)$ **D** $(2x + 7)(3x - 4)$

18. a The point $(3, c)$ lies on the line $7x - 12y = 5$. Find c.

    b What is the equation of the line that is parallel to the line $y = 8 - 3x$ and which passes through the origin?

    c At what point does the line $y = 5 - x$ cut

        i the y–axis _____

        ii the x–axis? _____

19. Expand:

    a $(x + 3)(x + 5)$ _____

    b $4a(b + 2c)$ _____

    c $(2a + b)(c + 2d)$ _____

    d $(3x - y)(3 - z)$ _____

20. Factorise:

    a $x^2 + 8x + 7$ _____

    b $x^2 + 2x - 35$ _____

    c $a^2 - 10a - 24$ _____

    d $p^2 - 49$ _____

21. a The average mass of the 11 players in a school cricket team is 56.8 kg. The average mass of the team including the 12th boy is 56.3 kg. Work out the mass of the 12th boy.

b How long will it take to travel 112 miles at 32 m.p.h.?

c How far will a car travel in 40 minutes at 96 km/h?

22.

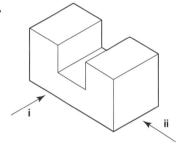

This solid is built up of 10 cubes of side 2 cm. Draw an accurate plan and elevations from the directions indicated by the arrows. Each square represents 1 cm.

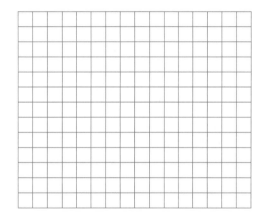

23.

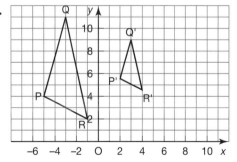

Draw PP', QQ' and RR' and continue all three lines until they meet. Write down:

a the coordinates of the centre of enlargement

_____

b the scale factor.

_____

24.

a Draw △ABC with A(3, 5), B(7, 4) and C(6, 7).

b i Draw △A'B'C', the enlargement of △ABC, using the point (5, 5) as the centre of enlargement and a scale factor of 2.

ii Write down the coordinates of A', B' and C'.

_____

iii Measure the lengths of the three sides of both triangles.

_____

iv Are the lengths of the three sides of △A'B'C' twice as long as the lengths of the corresponding sides of △ABC?

_____

c i Draw △A"B"C", the enlargement of △ABC, using the point (5, 5) as the centre of enlargement and a scale factor of –2.

ii Write down the coordinates of A", B" and C".

_____

iii Measure the lengths of the three sides of both triangles.

_____

iv Are the lengths of the three sides of △A"B"C" twice as long as the lengths of the corresponding sides of △ABC?

_____

15 Ratio and Proportion

1. Give the following ratios in their simplest form:

   a 24:16 _____

   b 64:72 _____

   c $\frac{1}{6} : \frac{2}{3} : \frac{1}{2}$ _____

   d 81:54:135 _____

   e $3\frac{2}{3} : 5\frac{1}{2}$ _____

2. Which is the larger ratio?

   a 20:7 or 23:8 _____

   b 3:11 or 5:22 _____

3. Express the following ratios in the form $n:1$ giving n correct to 3 s.f. where necessary:

   a 5:4 _____

   b 10:3 _____

   c 6:7 _____

4. Simplify the following ratios:

   a 53c:$1.59 _____

   b 35cm:0.2m _____

   c 675 mg:1 g _____

5. Find the ratio of the following prices:

   a $84 for 12 to $8 each

   _____

   b $50 per kg to $40 000 per tonne

   _____

   c $2700 per metre to $36 per cm.

   _____

6. A rectangle is 12 cm long and 8 cm wide. A second rectangle is 8 cm long and 5 cm wide. Find the ratio of their:

   a lengths _____

   b widths _____

   c perimeters _____

   d areas. _____

7.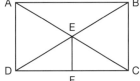

   ABCD is a rectangle measuring 12 cm by 6 cm. The diagonals of the rectangle cross at E, and F is the midpoint of DC.

   Find the ratio of the following areas:

   a △ABC:△ABE _____

   b △DEC:△EFC _____

   c △DEF:△DBC _____

   d △DEF:BCFE _____

8.

   The diagonals divide this square into regions which are marked with letters. Find the ratios of the areas marked with following letters:

   a A:B _____

   b A:C _____

   c D:A + B _____

   d C:D + E _____

9. Find the value of x if:

a $x:3 = 5:4$ _____

b $x:6 = 3:7$ _____

c $4:x = 2:5$ _____

d $5:6 = x:2$ _____

10. The ratio of the number of girls to the number of boys in a class is $7:5$. There are 15 boys. How many girls are there?

_____

11. a Divide $56 into two parts in the ratio $5:2$.

_____

b Divide 135 m into three parts in the ratio $1:3:5$.

_____

c Divide 1 hour 17 minutes into three parts in the ratio $2:3:6$.

_____

12. One litre of fuel takes a car 22 km. At the same rate how far does this car travel on:

a 4 litres _____

b 5.6 litres. _____

13. The cost of 1 kg of mixed vegetables is $264. Find the cost of $\frac{3}{4}$ kg.

_____

14. 6 cups and saucers cost $1920. What is the cost of one cup and saucer?

_____

15. The cost of running an electric fan for 4.5 hours is $23.40. What is the cost of running the fan for 1 hour?

_____

16. A machine uses 6 units of electricity in 4 hours. How many units does it use in 5 hours?

_____

17. A $\frac{1}{2}$ kg bag of sweets cost $600. At the same rate what would a $1\frac{3}{4}$ kg bag cost?

_____

18. Pamela changed EC$55 into US dollars and got $25 for them. How many US dollars would she get for EC$616?

_____

19. It cost $132 000 for tickets for a group of 12 students to attend a concert. How much would it cost for tickets for 19 students?

_____

20. An 8 kg bag of potatoes cost $336. At the same rate, what would a 50 kg bag cost?

_____

21. A school allows 95 exercise books a year for every 5 students. How many exercise books are needed for 24 students for a year?

_____

22. A recipe for Chinese Bean Sprouts to serve four lists the following ingredients:

50 g chicken	20 ml peanut oil
4 spring onions	500 g bean sprouts
root ginger	80 ml chicken stock
4 celery sticks	10 ml soy sauce
100 g mushrooms.	

a How much will the peanut oil cost if it is sold in litre bottles at $200 each?

_____

b Chicken costs $224 per kg. How much will the chicken cost?

_____

c A bottle of soy sauce contains 150 ml. How many servings of Chinese Bean Sprouts should this bottle be sufficient for?

_____

d List the ingredients to serve 10 people.

_____

_____

_____

_____

23. Kevin buys enough turf to lay a rectangular lawn measuring 36 m by 24 m. Before laying the turf he changes his mind. He decides that his rectangular lawn will be 32 m long. If he lays all the turf how wide is it?

_____

24. A short story is 121 lines with an average of 15 words per line. It is retyped with an average of 11 words per line. How many lines will there be?

_____

25. A spreadsheet containing the results of a survey has 34 rows with 12 cells in each row. The same results can be entered in the same number of cells but with 24 rows. How many cells are needed in each row?

_____

26. In a factory 63 machines are needed to produce the required number of units in 48 hours. How many machines are needed to produce the same number of units in 42 hours?

_____

27. In a large company 30 offices are needed if the staff are accommodated 8 to an office. How many offices would be needed if they are rearranged to accommodate 10 to an office?

_____

28. A book is 156 pages long if the text is arranged with 39 lines to each page. How many pages will be required if the text is re-set with the same size type but with 36 lines to a page?

_____

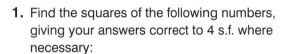

1. Find the squares of the following numbers, giving your answers correct to 4 s.f. where necessary:

   a 4.9 _____

   b 8.3 _____

   c 34.7 _____

   d 0.7245 _____

   e 0.0044 _____

   f 78 _____

2. Find the square roots of the following numbers, giving your answers correct to 4 s.f. where necessary:

   a 8.46 _____

   b 1.93 _____

   c 0.075 _____

   d 0.4 _____

   e 0.000646 _____

   f 54.3 _____

In the remaining questions give lengths that are not exact correct to 3 s.f.

3. Find the length of AC.

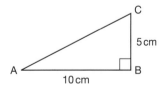

_____

4. Find the length of PR.

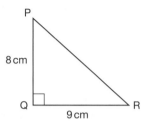

_____

5. Find the length of XZ.

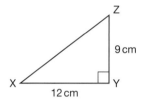

_____

6.
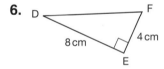

Find the length of DF.

_____

7. In $\triangle$ABC, $\hat{B} = 90°$, AB = 9 cm and BC = 11 cm. Find the length of AC.

_____

8. In $\triangle$XYZ, $\hat{Y} = 90°$, XY = 10 cm and YZ = 3 cm. Find the length of XZ.

_____

In questions 9 to 11 find the length of the hypotenuse in each right-angled triangle.

9.

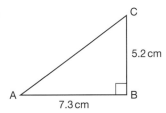

b

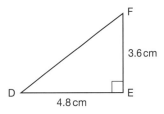

10.

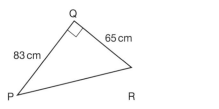

c

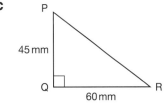

11.

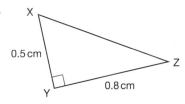

15. Find the length of BC.

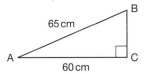

12. In △ABC, B̂ = 90°, AB = 6.4 cm and BC = 8.7 cm. Find the length of AC.

16. Find the length of DE.

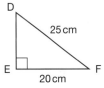

13. In △XYZ, Ŷ = 90°, XY = 12.4 cm and YZ = 16.8 cm. Find the length of XZ.

14. In this question decide whether each triangle is similar to a 3, 4, 5 triangle or to a 5, 12, 13 triangle, or neither. Find the length of hypotenuse using the method you think is easiest.

17. Find the length of PQ.

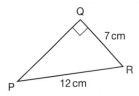

a

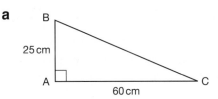

18. Find the length of YZ

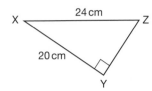

_____

19. a In △ABC, $\hat{B} = 90°$, AB = 5 cm and
AC = 7 cm. Find the length of BC.

_____

b In △PQR, $\hat{Q} = 90°$, PQ = 2.25 cm
QR = 7.75 cm. Find the length of PR.

_____

c In △XYZ, $\hat{Y} = 90°$, XY = 6.83 cm and
XZ = 9.37 cm. Find the length of YZ.

_____

20. a In △ABC, $\hat{C} = 90°$, AB = 76 cm and
BC = 32 cm. Find the length of AC.

_____

b In △PQR, $\hat{P} = 90°$, PQ = 4.8 cm QR =
7.3 cm. Find the length of PR.

_____

c In △XYZ, $\hat{X} = 90°$, XY = 5.27 cm and
XZ = 2.45 cm. Find the length of YZ.

_____

21. AB = BC = 12 cm.
AC = 16 cm.
Find the height of
the triangle.

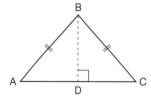

_____

22. In △PQR, PQ = QR = 12 cm and the
perpendicular height from Q to PR is 9 cm.
Find the length of PR.

_____

23. A circle with centre O has a radius of 6 cm. A
chord AB is of length 9.6 cm. Find the distance
of this chord from the centre of the circle.

_____

24. In a circle with centre O and radius 6.4 cm, a
chord AB is 3.7 cm from O. Find the length of
the chord.

_____

25. a Find the length of the diagonal of a square
of side 15 cm.

_____

b Find the length of the diagonal of a
rectangle measuring 15 cm by 10 cm.

_____

26. A diagonal of a rectangular lawn is 24 m long.
Each long side measures 18 m. How wide is
the lawn?

_____

27. The diagram shows the side view of a
workshop. Find the slant length of the roof.

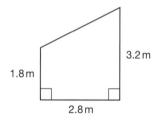

_____

28. A ship sails 56 nautical miles due south, then
48 nautical miles due east. How far is it from
its starting point?

_____

29. The length of the diagonal of a square is
18 cm. What is the length of a side of this
square?

_____

30.

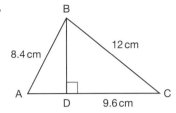

In the diagram BDC = 90°, AB = 8.4 cm, BC = 12 cm and CD = 9.6 cm.

Find: **a** the length of BD _____

  b the length of AC. _____

Is AB̂C = 90°?

Give a reason for your answer.

_____

31. Are the following triangles right-angled?

  a Triangle ABC: AB = 4.8 cm, BC = 3.6 cm, AC = 6 cm.

_____

  b Triangle PQR: PQ = 9 cm, QR = 20 cm, PR = 22 cm.

_____

  c Triangle XYZ, XY = 7.2 cm, XZ = 3 cm, YZ = 7.8 cm.

_____

32. Find the missing lengths in the following triangles:

  a

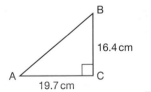

_____

  b

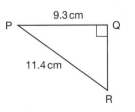

_____

c

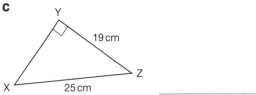

_____

33. a In △PQR Q̂ = 90°, PQ = 47 cm and QR = 35 cm. Find the length of PR.

_____

  b In △DEF, F̂ = 90°, DF = 3.7 cm and DE = 4.8 cm. Find the length of EF.

_____

  c In △XYZ, XY = 56 cm YZ = 65 cm and XZ = 33 cm. Show that the triangle is right-angled and state which angle is 90°.

_____

34.

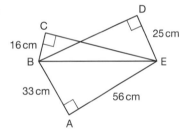

In the diagram triangles ABE, BCE and BDE are right-angled. Find the length of:

  a BE _____

  b CE _____

  c BD. _____

Give all angles correct to 1 d.p. and all lengths correct to 3 s.f.

1. Find the tangents of the following angles:

   a 24° _____

   b 72° _____

   c 39° _____

   d 68.8° _____

   e 26.4° _____

   f 7.6° _____

2. Find the angle whose tangent is

   a 0.7743 _____

   b 1.534 _____

   c 0.2597 _____

   d 0.4253 _____

   e 3.2 _____

   f 0.3592 _____

3.

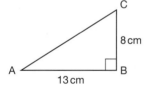

Find:

   a tan A _____

   b tan C. _____

4. In $\triangle XYZ$, $\hat{Y} = 90°$, $XY = 7.4$ cm and $YZ = 8.8$ cm. Find:

   a the size of $\hat{X}$ _____

   b the size of $\hat{Z}$. _____

5.

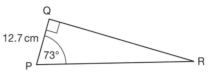

Find the length of QR _____

6.

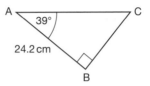

Find the length of BC. _____

7. In $\triangle XYZ$, $\hat{Z} = 90°$, $\hat{Y} = 34°$ and $YZ = 8.3$ cm. Find the length of XZ.

_____

8.

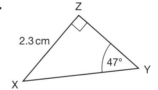

Find the length of YZ. _____

9.

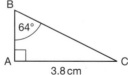

Find the length of AB. _____

10. In $\triangle XYZ$, $\hat{Z} = 90°$, $\hat{Y} = 42°$ and $XZ = 15$ cm. Find the length of YZ.

_____

11. In $\triangle ABC$, $\hat{C} = 90°$, $\hat{A} = 53.4°$ and $BC = 4.8$ cm. Find the length of AC.

_____

12. Find the size of Â.

a

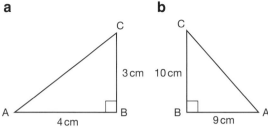

b

Â _____

Â _____

c

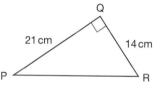

Â _____

13.

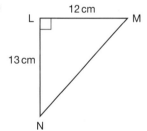

a Find the size of:

i P̂ _____

ii R̂. _____

b

L 12 cm M

13 cm

N

Find the size of:

i M̂ _____

ii N̂. _____

14. Find the sines of the following angles:

a 43.7° _____

b 82.6° _____

c 13.4° _____

d 51.8° _____

15. Find the angle whose sine is

a 0.3417 _____

b 0.9384 _____

c 0.721 _____

d 0.4545 _____

16.

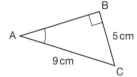

Find the size of Â. _____

17.

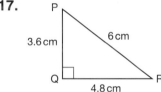

a Find the size of P̂ _____

b Find the size of R̂. _____

In questions 18 and 19 use the information in the diagram to find the required length.

18.

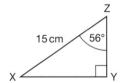

Find the length of XY. _____

19.

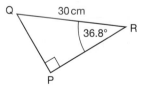

Find the length of PQ. _____

20. In △ABC, Â = 90°, BC = 16 cm and Ĉ = 42°. Find the length of AB.

_____

21. In △XYZ, X̂ = 90°, YZ = 7.4 cm and Ŷ = 34°. Find the length of XZ.

_____

22. Find the cosines of the following angles:

a 12.4° _____

b 83.9° _____

c 56.2° _____

d 16.8° _____

e 39.2° _____

f 72.7° _____

23. Find the angle whose cosine is

a 0.542 _____

b 0.7164 _____

c 0.8843 _____

d 0.5754 _____

e 0.9218 _____

f 0.2189 _____

In questions 24 and 25 find the size of the marked angle.

24.

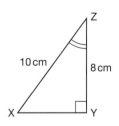

_____

25.

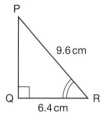

_____

26. In △PQR, P̂ = 90°, QR = 14 cm and PQ = 5.8 cm. Find the size of Q̂.

_____

27. In △DEF, F̂ = 90°, DE = 24 cm and DF = 18.4 cm. Find the size of D̂.

_____

In questions 28 and 29 find the required length.

28.

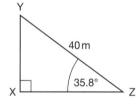

Find the length of XZ. _____

29.

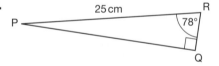

Find the length of RQ. _____

30. In △ABC, B̂ = 90°, Ĉ = 23° and AC = 12 cm. Find the length of BC.

_____

In questions 31 to 33 find the marked angles.

31.

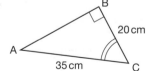

_____

32.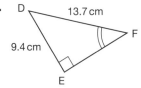

33.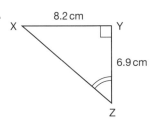

34. In △PQR, R̂ = 90°, PQ = 18 cm and QR = 8 cm. Find the size of P̂.

35. In △DEF, Ê = 90°, DE = 3.6 cm and FE = 4.7 cm. Find the size of F̂.

36. In △ABC, B̂ = 90°, Ĉ = 23° and AC = 12 cm. Find the length of BC.

In questions 37 to 39 find the required length.

37.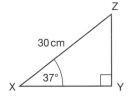

Find the length of YZ. _____

38.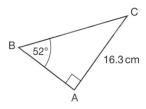

Find the length of AB. _____

39.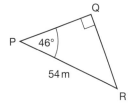

Find the length of PQ. _____

40. In △ABC, Ĉ = 90°, Â = 67° and AC = 30 cm. Find the length of BC.

41. In △PQR, P̂ = 90°, Q̂ = 31° and RQ = 42 cm. Find the length of PQ.

In questions 42 and 43 find the length of the hypotenuse.

42.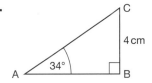

43.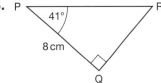

44. In △PQR, P̂ = 90°, Q̂ = 40° and PR = 9 cm. Find the length of QR

45. In △XYZ, Ẑ = 90°, Ŷ = 73° and YZ = 30 cm. Find the length of XY.

1. From a point on level ground 50 m from the base of a building, the angle of elevation of the top of the building is 42°. Find the height of the building.

_____

2. A is the point (3, 0) and B is (10, 6). Calculate the angle between AB and the x–axis.

_____

3. P is the point (2, 2), Q is (2, 8) and R is (7, 8). Calculate angle PRQ.

_____

In questions 4 to 6, A is a point on the ground and Â is the angle of elevation of C, the top of BC. Find the height of BC.

4.

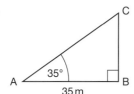

BC = _____

5.

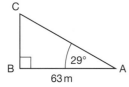

BC = _____

6.

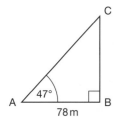

BC = _____

7. A tower is 82 m high. From a point P on the ground, the angle of elevation of the top of the tower is 38°. Find the distance from P to the foot of the block.

_____

8.

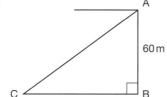

From a point A, at the top of a vertical tower 60 m high, the angle of depression of a milestone C on the ground below is 37°. Find the distance of C from the foot of the tower.

_____

9. From the top of a cliff, which is 70 m high, the angle of depression of a yacht out at sea is 35°. How far is the yacht from the base of the cliff?

_____

10. Viewed from the top of a cliff 54 m high, the angles of depression of two boats directly out to sea are 44° and 32°. Find:

    a the distance from each boat to the base of the cliff

    _____

    b the distance between the boats.

    _____

11.

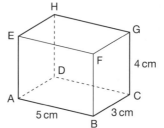

a Which edges of this cuboid are equal in length to AB?

_____

b Which edges of this cuboid are equal in length to BC?

_____

c Join A to F.

   i Use Pythagoras' theorem to find the length of AF.

_____

   ii Find the size of BÂF by using its tangent.

_____

d Join F to C. Find FĈB.

_____

e Join BD and BH. Find:

   i the length of DB

_____

   ii the size of HB̂D.

_____

12.

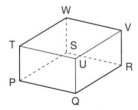

In this cuboid PQ = 12 cm, PS = 20 cm and PT = 5 cm. Find:

a the length of QS _____

b the length of QW _____

c the size of WQ̂S. _____

13.

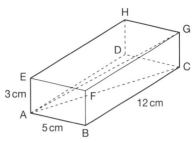

In the cuboid AB = 5 cm, AE = 3 cm and BC = 12 cm. Find:

a the length of AC _____

b the size of CÂG _____

c the length of AG. _____

Use this cuboid to answer questions 14 and 15.

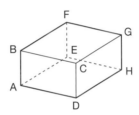

14. In the cuboid AB = 5 cm and AD = 8 cm. Find:

a the length of BD _____

b the size of the angle between AD and BD.

_____

15. In the cuboid AB = 10 cm, AD = 15 cm and AE = 20 cm. Find:

a the length of AH _____

b the length of AG _____

c AĤE _____

d AĤB _____

e GÂH. _____

In questions 16 to 20 draw a rough sketch outside your workbook to show each of the following bearings. Mark the angle in your sketch.

16. From a point A the bearing of my school, P, is 135°.

17. The bearing of the Princess Hotel, H, from a point B is 056°.

18. From a point P the bearing of the cricket ground, G, is 263°.

19. The bearing of Peke's restaurant, P, from the town hall, H, is 320°.

20. From a village, A, the bearing of another village, B, is 125°.

21. D is 20 km due north of E. The bearing of F from D is 146° and the bearing of F from E is 056°. Find:

a the distance of E from F

b the bearing of D from F.

22. A ship sails 8 km on a bearing of 042°, then 10 km on a bearing of 132°. How far is the ship from its starting point?

23. Q is 30 km due east of P. The bearing of R from P is 061° and the bearing of R from Q is 331°. Find:

a the angles of triangle PQR

b the distance PR.

24. From a ship, P, a lighthouse, L, is 600 m away on a bearing of 038°, and a yacht, Y, is 780 m away on a bearing of 132°. Find the distance and bearing of the lighthouse from the yacht.

19 Probability and Statistics

1. A letter is chosen at random from the letters in the word CONDOMINIUM. Find the probability that the letter is

   a N _____

   b M _____

   c D _____

   d a vowel _____

   e not a vowel. _____

2. One card is taken at random from an ordinary pack of 52 playing cards. Find the probability that it

   a is a king

   _____

   b has a face value of 3, 4 or 5

   _____

   c is not black

   _____

   d is a king, a queen or a jack.

   _____

3. a Write down the probability of throwing a 3 with an ordinary fair die.

   _____

   b A die is thrown 90 times. How many times is the score likely to be 3?

   _____

4. Find the mean, mode and median of each set of numbers:

   a 10, 8, 12, 15, 14, 13, 12

   Mean _____

   Mode _____

   Median _____

   b 1.4, 1.8, 1.7, 1.2, 1.2, 1.3, 1.2

   Mean _____

   Mode _____

   Median _____

   c 2.8, 1.7, 2.7, 2.5, 3.1, 2.9, 3.4, 2.1

   Mean _____

   Mode _____

   Median _____

5. A school entered 8 girls in a swimming competition. The marks they scored were:

   72, 95, 85, 43, 75, 82, 63, 59.

   Find:

   a the mean mark _____

   b the median mark. _____

   c Which of these two, the mean or the median, gives the best representation of the group as a whole? (Briefly say why the one you choose is better.)

   _____

6. Stuart counted the number of letters in the words in a paragraph of a book he was reading.
   They were:

2	5	7	5	2	8	8	7	6	3
4	7	6	12	1	7	13	9	9	8
5	9	4	6	6	9	3	10	12	7

   How many words were there in the paragraph? _____

   For the data find:

   a the mean number of letters per word

   _____

   b the mode _____

   c the median. _____

7. The table shows the number of tickets bought per person for a pop concert by the first group of people in a queue.

No. of tickets bought	1	2	3	4	5	6	8
Frequency	300	250	120	45	3	10	2

    a How many of these people bought tickets?

    _____

    b Find the mean number of tickets bought.

    _____

    c Find the mode. _____

8. This table shows the number of plants bought by the customers at a garden centre.

Number of plants	1	2	3	4	5	6
Frequency	15	21	8	5	3	1

Find:

    a the mean number of plants bought

    _____

    b the modal number. _____

9. Four coins were tossed together 40 times and the number of heads per throw was recorded in a table.

No. of heads	0	1	2	3	4
Frequency	3	10	17	8	2

Find:

    a the median number of heads per throw

    _____

    b the mode _____

    c the mean. _____

10. Dr Ali recorded the number of patients he saw each hour over a period of a week. The data is given in the table.

Number of patients	4	5	6	7	8	9	10
Frequency	5	8	12	8	7	3	4

    a How many patients did he see?

    _____

    b For the data find:

       i the median _____

       ii the mode _____

       iii the mean. _____

11. A test was marked out of 30. The scores of the students who took the test are given below:

15   6 18 13 10 24 26 16

12 23 14   8 12 16 23 12

14 25 21 24 18   7 18 21

    a Organise these scores in a grouped frequency table. Use groups 1–10, 11–20, 21–30.

Score	Frequency

    b How many students had a score less than 21?

    _____

    c Find the mean score.

    _____

12. The table shows the masses, to the nearest gram, of some apples.

Mass (g)	Frequency
1–50	18
51–100	42
101–150	53
151–200	32
201–250	9

a How many apples were weighed?

_____

b i How many apples weighed more than 100 g?

_____

ii What proportion was this?

_____

c Find the mean mass.

_____

d Use graph paper to draw a histogram to illustrate this data.

13. The amounts of money Mrs Ramon collected for her favourite charity are given in this frequency table.

Amount ($)	1–50	51–100	101–150	151–200
Frequency	12	54	69	15

a How many people contributed towards Mrs Ramon's favourite charity?

_____

b Find the mean amount of money contributed.

_____

c Use graph paper to draw a histogram to illustrate this data.

14. The table shows the times, to the nearest minute, that Dania had to wait for a bus each morning.

Time (minutes)	1–5	6–10	11–15	16–20
Frequency	9	14	5	2

a Find the mean number of minutes Dania had to wait for a bus.

_____

b Use graph paper to draw a histogram to illustrate this data.

15.

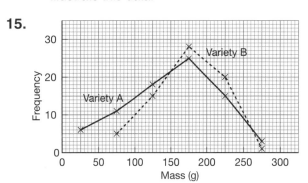

The diagram illustrates the distributions of two varieties of sweet potatoes.

a Which variety yielded the heavier potatoes?

_____

b How many potatoes were weighed for variety A?

_____

c How many potatoes were weighed for variety B?

_____

d Just from looking at the graph, which variety would you expect to have the higher mean mass?

_____

e i Work out the mean mass for each variety.

_____

ii Does your answer agree with the answer you gave for part (d)?

_____

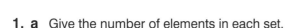

1. a Give the number of elements in each set.

    i {vowels}

_____

    ii {players in basket ball team}

_____

  b C = {prime numbers between 20 and 35}

    i Find $n\{C\}$

_____

    ii Write down the subset of C whose elements are all the odd numbers.

_____

    iii Is this a proper subset of C?

_____

    iv Write down the subset of C whose elements are even numbers.

_____

    What special name is given to this set?

_____

2. List the members of the set

  a $\{x: 5 \leqslant x \leqslant 12, x \in \mathbb{N}\}$

_____

  b $\{x: -3 < x < 4, x \in \mathbb{Z}\}$

_____

3. $U = \{x: -5 \leqslant x < 10, x \in \mathbb{Z}\}$
A = {even numbers}
B = {negative numbers}
C = {positive prime numbers}

List the sets A, B and C

_____

_____

_____

4. Give the complements of the following sets:

  a A = {2, 3, 5, 7} if U = {1, 2, 3, 4, 5, 6, 7, 8}

_____

  b C = {American cities} if U = {cities of the world}

    (You do not need to name any city.)

_____

  c P = {parallelograms} if U = {quadrilaterals}

_____

5. a If A = {people with a cell phone} and A' = {people without a cell phone} what is U?

_____

  b If P = {women who own a car} and P' = {women who do not own a car} what is U?

_____

6. U = {1, 2, 3, 4, 5, 6, 7, 8, 9, 10, 11, 12}
A = {multiples of 3}
B = {even numbers}

Show these on a Venn diagram and use it to list the sets:

A' _____

B' _____

$A \cup B$ _____

$(A \cup B)'$ _____

$A' \cap B'$ _____

7. U = {different letters in the word DICTIONARY}
A = {different letters in the word NATION}
B = {different letters in the word DAINTY}

Show U, A and B on a Venn diagram. Hence list the sets:

A' _____

B' _____

$A \cup B$ _____

$(A \cup B)'$ _____

$A' \cup B'$ _____

$A' \cap B'$ _____

8.

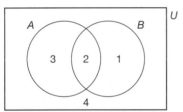

In this Venn diagram the numbers in the various regions show the number of elements or members of the set in that region. Find:

$n(A)$ _____

$n(B)$ _____

$n(A \cup B)$ _____

$n(A' \cup B')$ _____

$n(A \cap B)'$ _____

9.

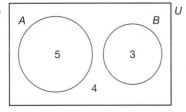

In this Venn diagram the numbers in the various regions show the number of elements or members of the set in that region. Find:

$n(A)$ _____

$n(B)$ _____

$n(A \cup B)$ _____

$n(A' \cup B')$ _____

$n(A \cap B)'$ _____

10.

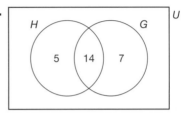

The Venn diagram shows the number of boys in a class who study history (H), and how many study geography (G). How many boys study:

a both subjects _____

b only history _____

c geography _____

d just one of these subjects? _____

11.

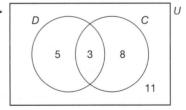

The Venn diagram shows the number of students in a class who had a dog (D) and how many had a cat (C). How many students:

a were there in the class _____

b did not have a dog _____

c had more than one pet _____

d had exactly one pet _____

e had at least one pet? _____

12. Shoppers leaving a market stall were questioned about what they had bought. 15 had bought oranges, 7 had bought mangoes, 4 had bought both and 11 had bought neither of these fruits. Show this information on a Venn diagram.

14.

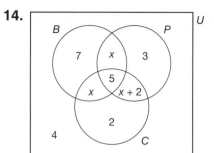

The Venn diagram shows the number of students in a class of 35 who study one or more of the sciences biology (B), chemistry (C), physics (P) or none of these.

a How many students do not study a science?

_____

b How many students study just one of these sciences?

_____

c Write down, in terms of x, an expression for the number of students who are studying one or more of these sciences. Hence or otherwise, find the value of x.

_____

13.

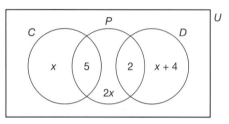

The Venn diagram shows the number of students taking chemistry (C), physics (P) and drama (D) in a class of 35. Every student studied at least one of these subjects.

a Write down, in terms of x, an expression for the number of students who take physics.

_____

b Write down, in terms of x, an equation which shows all the information given in the Venn diagram.

_____

c Work out the number of students who take physics only.

_____

d Work out the number of students who take drama.

_____

e Work out the number of students who take just one of these subjects.

_____

15. The universal set $U =$

$\{1, 2, 3, 4, 5, 6, 7, 8, 9, 10\}$

$A = \{2, 4, 6, 8\}$

$B = \{1, 3, 5, 7\}$

$C = \{3, 6, 9\}$

a List the elements in the set $(A \cup B) \cap C$

_____

b Write down the value of $n(A \cup C)$

_____

c List the elements in the set $A \cup (B \cap C)$

_____

16.

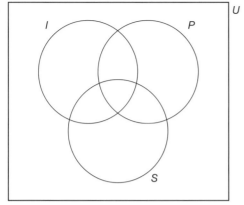

There are 8 people A, B, C, D, E, F, G and H, all of whom can speak at least one of the languages Italian (*I*), Portuguese (*P*) and Spanish (*S*).

A and B speak all three.

Only A, B and F can speak both Italian and Portuguese.

G and H speak Spanish only.

C and E speak Italian.

C speaks two languages.

No one except A and B speaks Portuguese and Spanish.

Only 4 people speak Portuguese.

Put the letters A, B, C, D, E, F, G and H in the correct regions.

17.

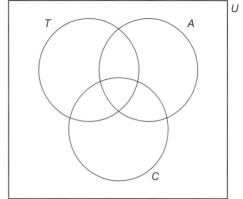

A questionnaire on 'Holiday Transport' given to 60 people produced the following facts: 40 had used a car (*C*), 28 had used an aeroplane (*A*), 16 had used a train (*T*), 18 had used both a car and an aeroplane, 5 had used both a car and a train, 4 had used only a train, 20 had used only a car and 6 had not been on holiday.

In the Venn diagram mark *x* as the number who had used all three forms of transport.

Calculate:

a the value of *x*

_____

b the number who went on holiday but did not use a car

_____

c the number who used more than one form of transport.

_____

1. a The mass of Xavier's luggage is more than the mass of Cath's luggage. Write an inequality to illustrate this.

_____

b Diana wants a lawn that is more than 20 m² but less than 30 m² in area. Write down two inequalities to illustrate this.

_____

_____

c Bill collects pens and pencils. He wants to have at least twice as many pens as he has pencils and at least 25 writing implements altogether. Illustrate these statements with two inequalities.

_____

_____

2. Use a number line to illustrate the range of values of x for which each of the following inequalities is true:

a $x > 5$

_____

b $x < -5$

_____

c $x > -3$

_____

d $x < \frac{1}{2}$

_____

3. Which of the inequalities given in question 2 are satisfied by a value of x equal to:

a 2 _____ **b** −4 _____

c 0 _____ **d** 0.3 _____

e −1.7 _____

4. Solve the following inequalities and illustrate your solution on a number line.

a $x - 3 > 5$

_____

b $x - 7 < 2$

_____

c $x + 3 < 0$

_____

d $x + 8 < 4$

_____

e $5x - 2 < x + 7$

_____

f $10 \leqslant 3 - 5x$

_____

5. On graph paper, draw diagrams to illustrate the following inequalities:

a $x < -3$

b $y \leqslant 2$

c $y > 2$

In questions 6 to 9 give the inequalities that define the *unshaded* regions:

6.

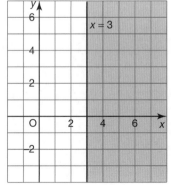

_____

7.

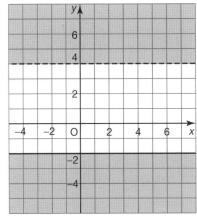

11. $3y \geqslant 4x$

12. $x - y < 3$

13. $2x - y \geqslant 4$

In questions 14 to 22 give the inequalities that define the *unshaded* regions.

14.

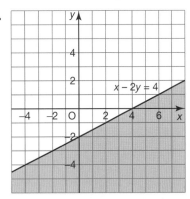

8.

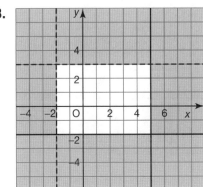

15.

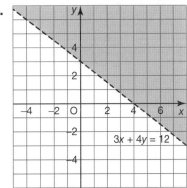

9.

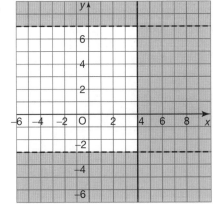

In questions 10 to 13 use graph paper and show the regions that illustrate the inequalities.

10. $y \leqslant x + 2$

16.

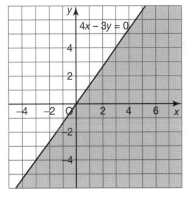

17.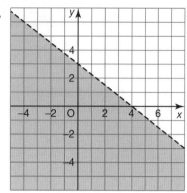

20.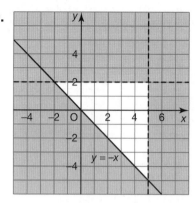

18.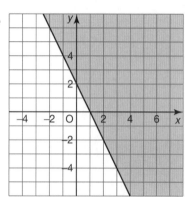

21.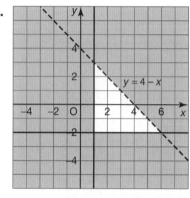

19.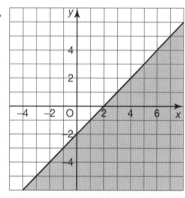

22.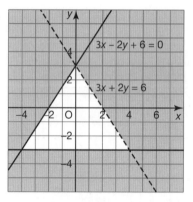

In questions 23 to 25 draw a diagram on graph paper to show the region defined by the inequalities. Leave the region unshaded.

23. $x > 0, y > 0, x + y < 5$

24. $x \leqslant 0, y \leqslant 1, x + y \geqslant -4$

25. $y < 4, x + y > 0, 2x - y < 0$

In questions 26 to 29 find the inequalities that define the *unshaded* region.

26.

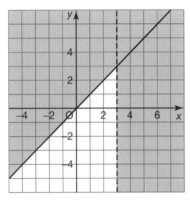

27.

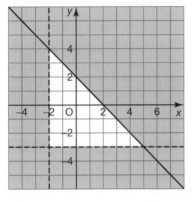

28.

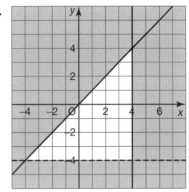

29.

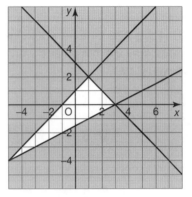

In questions 1 to 16 circle the letter for the correct answer.

1. In its simplest form the ratio 3.5 m to 145 mm is

   A 7:29   **B** 70:29

   C 700:29   **D** 7000:29

2. If 2.94 m is divided into two parts in the ratio 5:9, the length of the longer part is

   A 1.05 m   **B** 1.47 m

   C 1.68 m   **D** 1.89 m

3. A builder takes 5 days to build a wall 2 m high and 30 m long. The time he would take to build a similar wall 2.4 m high and with the same length would be

   A $5\frac{1}{2}$ days   **B** 6 days

   C $6\frac{1}{2}$ days   **D** 8 days

4. The square of 0.16 is

   A 0.0256   **B** 0.256

   C 0.4   **D** 0.04

5. The square root of 0.25 is

   A 0.05   **B** 0.0625

   C 0.5   **D** 0.625

6. In a right–angled triangle the length of the hypotenuse is 12 cm and the length of one of the other sides is 8 cm. The length of the remaining side is

   A 7 cm   **B** $\sqrt{80}$ cm

   C 8 cm   **D** 9 cm

7. Which of the three sets of numbers are measures of the sides of a right–angled triangle?

   i {9, 16, 25}

   ii {8, 10, 12}

   iii {9, 12, 15}

   A i and ii   **B** i and iii

   C ii and iii   **D** iii only

8. Which of the following values cannot be the value of the cosine of an angle?

   i 1.45

   ii 0.45

   iii 2.54

   A i only   **B** ii only

   C iii only   **D** i and iii only

9. From the top of a cliff 80 m high the angle of depression of a fishing boat out at sea is 32°. The distance of the boat from the base of the cliff, in metres, is

   A 80 tan 32°   **B** $\dfrac{80}{\tan 32°}$

   C 80 cos 32°   **D** 80 sin 32°

10. For the angle 45° which of these statements are true?

   i The tangent of the angle is greater than its sine.

   ii The sine and cosine have the same value.

   iii The sine is greater than the cosine.

   A i only   **B** ii only

   C iii only   **D** i and ii only

11. The bearing of A from B is 153°. The bearing of B from A is

   A 027°   **B** 117°

   C 297°   **D** 333°

12. The quantity of milk in each of five jugs is

   0.46 litres   0.59 litres   0.73 litres

   0.64 litres   0.83 litres

   The mean amount of milk in these jugs is

   A 0.65 litres   **B** 0.6 litres

   C 0.58 litres   **D** 0.53 litres

13. Given that $P = \{2, 5, 7, 9\}$ and $U = \{1, 2, 3, 4, 5, 6, 7, 8, 9, 10\}$, the complement of P is

   A {1, 3, 6, 8, 10}   **B** {1, 3, 4, 6, 10}

   C {1, 3, 4, 8, 10}   **D** {1, 3, 4, 6, 8, 10}

14. $U = \{1, 2, 3, 4, 5, 6, 7, 8, 9, 10, 11, 12\}$,
$P = \{$multiples of 3$\}$, $Q = \{$prime numbers$\}$

$(P \cup Q)'$ is

A $= \{3\}$

B $= \{2, 3, 5, 6, 7, 9, 11, 12\}$

C $= \{1, 4, 6, 8, 9, 10, 12\}$

D $= \{1, 4, 8, 10\}$

15. Which point satisfies the inequality
$x + 2y < 6$?

A (3, 2)   **B** (2, 3)

C (2, 2)   **D** (1, 2)

16. A letter is chosen at random from the word SEPTEMBER. The probability that it is a vowel is

A $\frac{1}{9}$ **B** $\frac{2}{9}$

C $\frac{1}{3}$ **D** $\frac{4}{9}$

17. a Which is the larger ratio, $15:7$ or $13:6$?

_____

b Divide \$48 into three parts in the ratio $2:3:7$.

_____

18. In a cinema there are 24 rows with 36 seats in each row. If these are rearranged in 27 rows how many seats will there be in each row?

_____

19. In △XYZ, $\hat{X} = 90°$, XZ = 5.4 cm and $\hat{Z} = 37°$. Find:

a the size of $\hat{Y}$ _____

b the length of XY   _____

c the length of YZ.   _____

20.

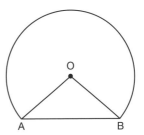

The diagram shows the cross-section through a tunnel. O is the centre of a circle. OA = OB = 7 m. The road, AB, is 12 m wide.

How high is the tunnel? _____

21. a

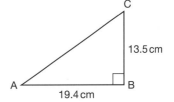

Find the length of AC. _____

b

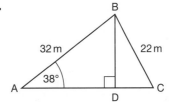

Find the length of QR. _____

22.

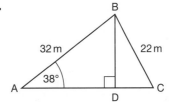

Find:

a the length of AD   _____

b the length of BD   _____

c the size of $D\hat{B}C$ _____

d the length of DC.   _____

e Is $A\hat{B}C$ a right angle?   _____

23.

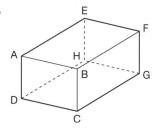

In the cuboid CD = 6 cm, CG = 9 cm and GF = 5 cm.

Find:

a the length of AF _____

b the size of DF̂A _____

c the length of AC _____

d the size of AĈD. _____

24. a The bearing of X from Y is 146°. What is the bearing of Y from X?

_____

b The bearing of P from Q is 218°. What is the bearing of Q from P?

_____

c From a point A, a point B is 5 km on a bearing of 042°. From B a third point C is 8 km on a bearing of 134°. Draw a sketch to show the relative positions of A, B and C. How far is C

i east of A _____

ii south of A? _____

25. The table shows the lengths, correct to the nearest cm, of 40 leaves from the same tree.

Length of leaf (cm)	5–9	10–14	15–19	20–24
Frequency	5	11	15	9

a Find the mean length of these leaves.

_____

b Draw a histogram to illustrate this data.

26.

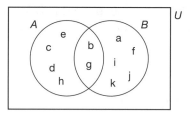

Use the information given in the Venn diagram to find:

a $n(A)$ _____

b $n(B)$ _____

c $n(A \cup B)$ _____

d $n(A \cap B)$ _____

e $n(A')$. _____

27.

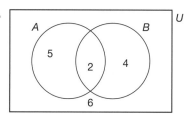

The numbers in the various regions of this Venn diagram show the number of elements in, or members of, the set in that region. Use this information to find:

a $n(A)$ _____

b $n(B)$ _____

c $n(A')$ _____

d $n(B')$ _____

e $n(A \cup B)$ _____

f $n(A \cap B)$ _____

g $n(A' \cup B')$ _____

h $n[(A \cap B)']$. _____

28.

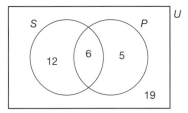

The Venn diagram shows how many families in a street have a chicken shed (*S*) and how many have a pool (*P*).

a How many houses are there in the street?

_____

b How many houses are there who do not have a pool?

_____

c How many have either a chicken shed or a pool but not both?

_____

29.

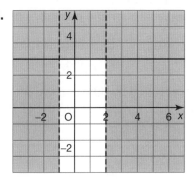

Find the inequalities that define the *unshaded* region.

_____

_____

30

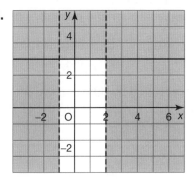

Find the inequalities that define the *unshaded* region.

_____

_____

Review Test 4: Units 1 to 21

In questions **1** to **16** circle the letter for the correct answer.

1. The value of $5 + 15 \div 5$ is

   A 3   **B** 4

   C 8   **D** 10

2. The correct symbol to place between the numbers 0.53 and $\frac{13}{15}$ is

   A <   **B** =

   C >   **D** ⩾

3. The value of $6 - 2 - 4(-2)$ is

   A 14   **B** 12

   C 4   **D** -4

4. Given that $p = \dfrac{q}{2 - r}$ the value of p when $q = 9$ and $r = \frac{1}{2}$ is

   A 4   **B** 5

   C 6   **D** 8

5. 30% of 50 exceeds 40% of 30 by

   A 10   **B** 6

   C 4   **D** 3

6. Which of the following points does not lie on the line with equation $y = 3x - 5$?

   A (2, 1)   **B** (-1, -2)

   C (3, 4)   **D** (0, -5)

7. The point $(4, -2)$ lies on the line $5x - 3y = 2n$. The value of n is

   A 26   **B** 14

   C 13   **D** 7

8. The expansion of $(2 - 3x)^2$ is

   A $4 - 12x - 9x^2$ **B** $4 + 12x - 9x^2$

   C $4 - 12x + 9x^2$ **D** $4 + 12x + 9x^2$

9. For the equation $y = 2x^2 - 3x - 4$ the value of y when $x = -2$ is

   A -4   **B** 6

   C 10   **D** 18

10.

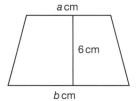

In this trapezium, the lengths of the parallel side are a cm and b cm. If the perpendicular distance between the parallel sides is 6 cm and the area of the trapezium is 54 cm^2 the value of $a + b$ is

   A 9   **B** 12

   C 16   **D** 18

11. The equation of the straight line with gradient $-\frac{1}{2}$ which passes through the point $(3, -2)$ is

   A $x - 2y + 1 = 0$ **B** $x + 2y + 1 = 0$

   C $x + 2y + 7 = 0$ **D** $x + 2y - 1 = 0$

12. A bicycle depreciated by 30% of its value during its first year. It was then worth $42 000. Its purchase price was

   A $54 600   **B** $60 000

   C $65 000   **D** $68 000

13. $3(5x - 2) - 4(3x - 2) =$

   A $3x - 14$ **B** $3x - 2$

   C $3x + 2$ **D** $3x + 14$

14.

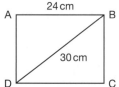

In this rectangle AB = 24 cm and BD = 30 cm. The perimeter of the rectangle, in centimetres, is

   A 42 cm   **B** 54 cm

   C 84 cm   **D** 104 cm

15. A quadrilateral with both pairs of opposite sides parallel is best described as a

A trapezium   **B** parallelogram

C kite   **D** cyclic quadrilateral

16. If $x \in A \cup B$ which of the following statements must be true?

A $x \in A \cap B$

B $x \in A$ or $x \in B$

C $x \in A$ and $x \in B$

D $x \in A$ and $x \in B$

17. a If x is an integer, find the values of x which satisfy

$x - 4 < 2x + 3 < 9.$

b Find the equation of the line through the point $(4, -3)$ which is parallel to the line $y = 3x - 7$.

18. a An article bought for $90 000 depreciates by 10% each year. What is it worth at the end of 2 years?

b An article bought for $9000 appreciates by 10% each year. What is it worth at the end of 2 years?

c An agent receives $90 000 commission on sales of $3 600 000. At what percentage rate was the commission paid to the agent?

19. a What sum of money, invested for 5 years at 4% p.a. simple interest, gives $60 000 interest?

b Find the compound interest on $500 000 for 2 years at 5%.

20. Factorise:

a $x^2 - 9x + 18 =$ _____

b $x^2 + 12x + 36 =$ _____

c $12 - x - x^2 =$ _____

d $6x^2 - 7x - 3 =$ _____

21.

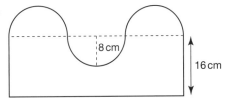

The curved edge of this shape is made up of three identical semicircles. Find:

a the perimeter of the shape

b its area.

22. Solve the equations

a $5 - 2(x - 3) = 1$

b $7 - 2x = 13 - 5x$

c $3(5x - 3) - 4(2x + 1) = 1$

23. Solve the simultaneous equations:

a $3x + 2y = 5$
   $x - y = 5$

b $4x - 5y = 14$
   $3x + 5y = -7$

c $4a + 3b = 9$
   $3a = 8 - b$

24.

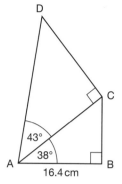

Use the information on the diagram to find:

a the length of AC

_____

b the length of AD

_____

c the length of DC

_____

d the area of triangle ACD.

_____

25.

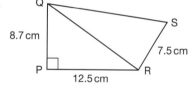

Use the information on the diagram to find:

a the size of PR̂Q _____

b the size of RQ _____

c the size of RQ̂S _____

26.

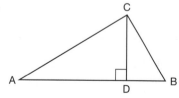

In this diagram the base, AB, of triangle ABC is 6 cm longer than its height, CD. The area of the triangle is 56 cm². Find the length of

a AB _____

b CD. _____

27. a For the first 60 km of a journey Mr Walcott travels at an average speed of 80 km/h. For the next 15 km he travels at an average speed of 20 km/h. Find his average speed for the whole journey.

_____

b The lengths of 6 carrots are

13.5 cm, 14 cm, 12 cm, 18 cm, 17.5 cm, 18 cm.

Find the mean, mode and median length of these carrots.

_____

_____

_____

28.

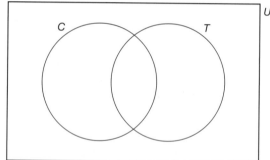

In a class of 30 girls, 16 like cricket (*C*), 13 like tennis (*T*) and 7 do not like either sport.

a In the Venn diagram fill in the correct numbers in the regions.

b Use your diagram to find how many girls

i like both cricket and tennis

_____

ii like either cricket or tennis but not both

_____

iii do not like tennis.

_____

29. This list gives the masses of 30 tomatoes.
The masses are in grams correct to the nearest gram.

142	220	164	98	65	137	151	156	192	174
147	87	113	182	235	163	183	48	183	112
102	190	172	83	137	76	159	134	59	95

a Complete this frequency table:

Mass (g)	Tally	Frequency
41–90		
91–140		
141–190		
191–240		

b In which group would a tomato with a mass of 189 g be recorded? _____

c Find the mean mass of these tomatoes. _____

30.

Time	Midnight	2	4	6	8	10	Noon	2	4	6	8	10	Midnight
Temp (°F)	12	11	9.5	10	19	35.5	60	78	68	34	22	15	13

The table shows the recorded temperatures in a town over a 24–hour period. On graph paper draw axes, using 2 cm = 2 hours on the horizontal (long) axis and 2 cm = 10°F on the vertical axis. Plot the points and draw a smooth curve through them.

Use your graph to estimate:

a the temperature at 9 a.m. and at 9 p.m. _____

b the times at which the temperature is 50°F. _____

c the fall in temperature between 5 p.m. and 5 a.m. _____

d the lowest temperature recorded. _____

31. Draw a diagram to show the region defined by the inequalities

$x > 1, y > 0, x + y < 5$

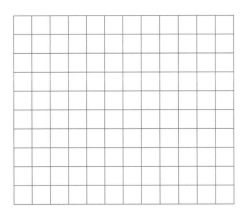